Anthropological Research Framing for Archaeological Geophysics

Anthropological Research Framing for Archaeological Geophysics

Material Signatures of Past Human Behavior

Jason Randall Thompson

ROWMAN & LITTLEFIELD
Lanham • Boulder • New York • London

Published by Lexington Books
An imprint of The Rowman & Littlefield Publishing Group, Inc.
4501 Forbes Boulevard, Suite 200, Lanham, Maryland 20706
www.rowman.com

Unit A, Whitacre Mews, 26-34 Stannary Street, London SE11 4AB

British Library Cataloguing in Publication Information Available

Library of Congress Cataloging-in-Publication Data Available

ISBN (cloth): 978-0-7391-7758-7
ISBN (electronic): 978-0-7391-7759-4

™ The paper used in this publication meets the minimum requirements of American National Standard for Information Sciences—Permanence of Paper for Printed Library Materials, ANSI/NISO Z39.48-1992.

Printed in the United States of America

Anthropological Research Framing for Archaeological Geophysics

Material Signatures of Past Human Behavior

JASON RANDALL THOMPSON

ROWMAN & LITTLEFIELD
Lanham • Boulder • New York • London

Published by Lexington Books
An imprint of The Rowman & Littlefield Publishing Group, Inc.
4501 Forbes Boulevard, Suite 200, Lanham, Maryland 20706
www.rowman.com

Unit A, Whitacre Mews, 26-34 Stannary Street, London SE11 4AB

British Library Cataloguing in Publication Information Available

Library of Congress Cataloging-in-Publication Data Available

ISBN (cloth): 978-0-7391-7758-7
ISBN (electronic): 978-0-7391-7759-4

∞™ The paper used in this publication meets the minimum requirements of American National Standard for Information Sciences—Permanence of Paper for Printed Library Materials, ANSI/NISO Z39.48-1992.

Printed in the United States of America

Contents

List of Illustrations

List of Tables

Acknowledgments

I WISH SINCERELY TO THANK DR. JAMES ENLOE OF THE UNIVERSITY OF IOWA, Dr. Glenn Storey of the University of Iowa, and Dr. Larry Conyers of the University of Denver. I owe all of you more than I can ever repay.

Thanks are also extended to Leah Rogers of Tallgrass Historians, LLC, Iowa City, Iowa.

Funding:

Research at Verberie in 2006 was funded by The University of Iowa Arts and Humanities Initiative Research Grant.

Research at Verberie in 2009 was funded by National Science Foundation Senior Research Grant BCS-0817575.

Introduction

THIS BOOK IS ABOUT DEVELOPING IDEAS FOR COLLABORATIVELY INTEGRATING anthropological archaeology and geophysical archaeology. The book is not strictly "about" archaeological geophysics (AGP) technology per se, nor is it a how-to guide for conducting archaeological research using such techniques. This is actually a book about one's own ideas for formulating anthropological/archaeological inquiries for engagement in a practice of collaborative "inquiry-driven archaeogeophysics" (Thompson et al. 2011). There will be repetition of certain themes throughout the work, in an attempt to address topics from multiple perspectives. The book is an invitation for increased dialogue between anthropological and geophysical archaeologists, so such repetition is used to make analyses contained therein as widely accessible as possible to the widest possible set of readers. In regards to inquiry-driven archaeogeophysics the questions we ask will be obviously be the drivers of our research. The quality and depth of our research will therefore be directly related to the quality and depth of our questions.

One assumes that you, the reader, are interested in the field of archaeological geophysics, or you wouldn't be reading this book. How, therefore, would you like to see the field develop or change, what changes would you make, and what would you be willing to do to help accomplish those things? Those will not be the last questions posed in this volume. Ideally, participation in innovating archaeological geophysics should be available to all or many interested parties. Innovation often comes from unexpected parties that have time and freedom (or funds and often luck, or even the lack thereof) to experiment or simply to ask new questions that hadn't yet been asked. Hence, the *extensive* use of questions in this book, not all of which will be answered. In fact, the book itself is largely dedicated to formulating useful anthropological research questions that we can, mutually, as anthropological archaeologists (AAs) and geophysical archaeologists (GAs) as a *collaborative* discipline of professionals, use geophysical technologies to address.

The book may also be to some degree controversial, in that the author wrote it with the intention of challenging both anthropological and geophysical archaeologists to be more mutually "co-relevant" and mutually collaborative for the betterment of our entire archaeological discipline in general.

By way of introducing the contents, chapter 1 consists of a general introduction and a summary of previous research on archaeological uses of geophysics and an overview of topics for this volume. Chapter 2 seeks to establish valid anthropological research frames or contexts for AGP research. The reason for this is that little such framing has even been attempted at this point despite advances in remote sensing and computing technology, and because of certain biases in publication patterns of AGP research. To date there has been very little substantive research into evaluating the anthropological significance of AGP, or even incorporation of elements of anthropological theory within AGP research design. The absence of theory in AGP leaves open many opportunities to integrate it now and in the future. Chapter 3 relates examples of biases. What might we wish to learn? What is possible for us to learn via AGP? What will we likely never learn? Chapter 4 describes research problems. Chapter 5 relates models, analogies, and how they are used in archaeology, as well as pattern recognition. Chapter 6 describes research framing. Chapter 7 details a very successful application of AGP at a French Upper Paleolithic open-air site (Verberie) and some of the humanistic anthropological inferences that are possible from it. It also describes how the field methodology was calibrated from the ground up through the use of controlled experiments on simulations in a sandbox and in an in-ground setting. Some key images will be presented, but only as many as absolutely necessary to illustrate points. While such images are unquestionably important (some might argue the most important aspect of AGP applications), this is not a book about images but ideas. Hopefully, one's writing will illustrate the concepts one tries to elucidate. Some tables are also included for presentation of ideas regarding the structure and organization of our queries. Chapter 8 describes an "unsuccessful" application of AGP at a site. Chapter 9 describes hypothesis-forming. Chapter 10 provides ideas for future research.

A HISTORICAL ANALOGY

Please bear with the author. This analogy is not directly applicable to AGP, but it is relevant and one hopes the reader will see the connection the author intends. In terms of history, the electric guitar was born in the middle years of the twentieth century, and its manufacture in the United States was dominated by two major luthier companies, Gibson and Fender, that made, respectively, the iconic Gibson Les Paul and Fender Telecaster and Stratocaster models. Innumerable hit songs by legions of famous bands and musicians were written using these instruments, which became fixed in the public consciousness as "the" electric guitars for nearly all genres of popular music. In fact, the mass-produced high-grade instruments themselves (i.e., the Stratocaster) *determined* the structure and sound of many hit songs, because guitar players adapted themselves to the playing mechanics and sounds of particular instruments. Even many custom-built models today are virtual copies of the vintage original Strat, the Tele, and the Paul, often using precision OEM specifications applied to different exotic woods and paints and personal tastes in hardware color and electronic (pickup) configurations unavailable to regular consumers. For decades, until the late 1970s and early 1980s, electric guitars were built primarily according to the templates of the Tele, the Strat, and the Paul, and the actual sounds recorded on albums featured a consistent basic stasis. Lots of different hit songs have basically exactly the same guitar sound, especially during the late 1960s and the first half of the 1970s, because the players were using the same basic equipment. There was a very slow general evolution in playing style but little change in instrument design. Furthermore, to get professional results in terms of playing technique and especially guitar sound, one was forced to use absurdly expensive (and rare) professional equipment, special effects, and elaborate studio microphone arrays. There were few home recording studios, modelers (amplifiers and signal-processing devices that use software-based emulation of actual professional amplifiers), or indeed really *any* user-friendly ways to access excellent guitar sounds at rational real-world prices. Not coincidentally, there were also many fewer garage-band home guitar players because the instruments themselves and the amplifiers were so cost-intensive.

If anyone wanted a certain specific guitar sound, then that required playing specific guitars built according to the established but "unstable orthodoxy" controlled by the dominant manufacturers along with boutique, high-end expensive amplifiers (Thompson 2012). But then, in the late 1970s, in Los Angeles, an autodidactic guitar player, who had started his forays into music with classical piano and drums instead of guitar, began tinkering with various guitar models to get a Les Paul sound from a Stratocaster body style guitar. This musician was Eddie Van Halen (EVH), neither a luthier, nor an electrical engineer, nor anything but a self-taught garage and bar band guitar player, but from him the "modern" rock guitar sound was born and the first several Van Halen albums turned the rock world on its head with his sound and "tone." Everything in rock guitar playing, recording, and performance was thereafter changed and millions of new guitar players noticed (and guitar and amplifier manufacturers noticed all the millions of new guitar fans and potential customers). The basics for this were nothing but one person having time, freedom, and the ability to tinker with the basics of the craft without official guidance or interference.

One doesn't have any literature to cite for this in standard APA style, but from their own recollection gained by reading countless EVH interviews in magazines, and watching him on TV, and living through it, the process was something like this: EVH wanted a Strat to sound like a Les Paul, so he pulled the bridge (rearmost) pickup from a Les Paul and routed out the body of a Stratocaster and fixed the Les Paul pickup within the Strat body. The symphony of guitar sounds this innovation gave EVH is duly captured on numerous platinum Van Halen albums and hit songs, and provoked revolutions in electric guitar playing, design, amplifier and signal processing, and public accessibility. EVH changed the way people approach electric guitar and brought it to the public. The guitar-playing public began making new demands of equipment manufacturers and stopped buying many of the old, familiar models. Millions of aspiring guitarists entered the marketplace of guitars, amplifiers, and effects devices.

In the 1980s, virtually all electric guitar manufacturers began making similar assembly-line instruments that arrived factory stock with modified Strat bodies containing modified Les Paul pickups. For most of the 1980s and 1990s, just about every major band one could see on MTV videos or live featured guitar players using different versions of EVH's original "Super-strat" design electric guitar, manufactured in bulk quantities for a market eager to try and replicate EVH's "sound" in homes, garages, and bars all over the planet. Marshall Amplification, a prominent amplifier manufacturer, revised many of their products during the same time period to allow for greater and "sweeter" distortion at manageable bedroom volumes. Since the 1990s, first Peavey, and now Fender have employed EVH as a sort of genius in residence to design various configurations of electric guitars, pickups, and especially tube amplifiers built to his specific requirements and suggestions, so that millions of rock guitar fans can now virtually mimic EVH's guitar tone in their own pursuits, bands, and bedrooms. EVH has probably made guitar and amplifier manufacturers who wanted to market his sound to the public more money than he's actually made from his own original recording sales. There's a lot more to EVH and Van Halen than one has described, but it is the basics above that caused them to think about him in relation to the premises of this volume. If one person could change the entire world of rock guitar so profoundly and pervasively, what innovations can archaeologists collaboratively produce by engaging in a deliberate effort to extract more anthropological content from archaeological geophysics and incorporate them more basically in our anthropological archaeology research methodologies? What steps would we need to follow in order to do this? Who wants to participate in such an endeavor? Who will be allowed to participate by their colleagues? How can we change the underlying assumptions made about AGP survey without changing the basic equipment? Can AGP equipment and techniques be made more "archaeology-friendly"? Can anthropological archaeology perhaps drive innovation in geophysical technology?

We can't always know when, where, why, from whom, or how innovations will occur. We can be fairly certain, however, that innovation won't occur in any field if new developments are ignored or if people are not open to new ideas. If innovations occur are they accepted and welcomed or are they regarded with hostility? Do mature fields peopled by professionals with diverse viewpoints accept proposed innovations or fight to maintain the status quo? How durable or entrenched is the status quo? Sometimes parties seek to preserve the status quo, even for perfectly valid reasons, and miss out on opportunities to advance their fields. Sometimes accomplished professionals don't take time to understand different approaches from new and different fields. Other times we might misunderstand or fail to understand proposed innovations, or perhaps only infrequently consult professionals from other disciplines. Finally, there are simply some of us who are hostile to innovation and fear it. All of us are guilty of the preceding at one time or other and in general such things can contribute to communication gaps.

Lack of understanding and lack of communication will be common refrains in this volume, and there will be multiple examples of communication failures provided as evidence. Among many such failures, for example, Conyers (2012) lists several varieties of pervasive communication failures common in archaeological geophysics: poor communication by anthropological archaeologists (hereinafter AAs) to geophysical archaeologists (hereinafter GAs) regarding detailed site conditions, and especially soil/sediment properties; the failure of AAs adequately to consult finished geophysical reports written on the very sites the AAs hired the GAs to survey; and a basic tension between perception and reality (common in Anthropology as the Ideal vs. the Real) in relation to AGP technology's actual capabilities. Yet, what communication failures on the part of GAs would anthropological archaeologists highlight? We don't hear much about those and we should ask why. Archaeological geophysics isn't only the purview of geophysicists.

Returning to the guitar analogy, arguably only two people, and maybe really even only one, actually and fundamentally changed rock guitar sound forever: Jimi Hendrix and Eddie Van Halen, and only EVH altered the basic underlying paradigm regarding guitar design, that one could rearrange components from different guitars to get a desired tone. The other one played a right-handed instrument left-handed, which drastically altered tone, sound, and timbre, but that was within the parameters of an established guitar design. Jimi Hendrix didn't redesign the electric guitar and he didn't popularize it to the extent that guitar and amplifier manufacturers changed their product lines to serve millions of new fans. Eddie Van Halen reimagined it, how it could sound, how it could be played. And none of that happened because of specific knowledge of sonics, wave dynamics, or even basic electronics. It happened because someone tinkered with the *structure* of an established orthodoxy and turned it on its head and made a new orthodoxy. One thinks this can be a lesson for us, if we are willing to be open-minded and interdisciplinary. We can learn from this. We can learn from each other. How can we use the preceding guitar analogy for AGP? What cognitive and theoretical tinkering can we do, and how can we facilitate greater communication between GAs and AAs?

Answering the preceding questions is what much of the rest of this volume will be devoted to doing. The life lessons we each bring to our research vary largely as a function of our individual experiences and interests, and in that vein one offered the analogy above for reasons that should become obvious below. One's own experience with archaeological geophysics extends over nearly the past ten years, after taking a course in geophysics. Prior to that, one had always regarded seismic stratigraphy of crustal plates and the Chicxulub impact structure to be highly interesting curiosities without imagining that near-surface geophysical methods had already been or ever would be developed that were of high utility to archaeology. The lack of archaeological geophysics awareness was nothing but one's own ignorance. As a student, one had been too much focused upon one's own (lithic and faunal) biases and those impressed by one's mentors and instructors. One knew that one needed to learn the method and theory of archaeology, and also that one had to master the peculiar jargon of the discipline. It never occurred to one that archaeologists could use radar, electricity, magnetism, or other physical principles to study what lies below. It should have been obvious since one had, of course, learned about photographic remote sensing as an undergraduate.

Discovery, however, and broadened awareness often prompt reflection upon previous biases and approaches, and one cannot overstress that geophysics will become *more* common in our field over time and we can expect it to elicit changes in our cognitive approaches. One's own cognitive approach to AGP has, for example, moved through several key phases of development, mainly with respect to integrating AGP in anthropological archaeology. These phases were 1) a shallow general acquaintance lacking substantive competence in any AGP technique also lacking any concept of integrating AGP in anthropological archaeology; 2) development of minimal competence (in one AGP technique) but lacking theory and method to integrate AGP in anthropological archaeology; to 3) working competence with theory and method for integrating AGP in anthropological archaeology. Future cognitive changes in archaeological reasoning are inevitable, as more archaeologists make use of

AGP and find novel ways and sites to apply the techniques. That suggests the foreseeable proximate future would be a wise time for AAs to begin acquainting themselves with AGP in earnest. What we will discuss below are some *possible* (not all, or most, but just some) ways of thinking about and using AGP anthropologically and bringing it to bear on humanistic issues and some practical examples. This volume is emphatically *not* a how-to manual, and it is important that the reader understand this fact.

One will also state that what they seek in this volume is to contribute to a revitalization of discourse between the various anthropological camps one describes or to help generate the beginnings of it, between AAs and GAs. Why do many American AAs (including CRM professionals) appear to be so ignorant of archaeological geophysics? It's easy to get overly compartmentalized in our discipline, especially as we focus on particular technical analytical fields. Such compartmentalization can also be exacerbated, however, if we hew to our own narrow ranges of interest and fail to challenge ourselves not only to stay abreast of innovations in our field but actively to learn about parallel research programs. Given the dynamics and pressures of an academic life, the variable Time is often what we lack in abundance, in that we often simply don't have enough time to keep track of all the things we should. AGP technologies are also often prohibitively expensive, results produced are (allegedly) inscrutable for many of our anthropological colleagues (Conyers 2012, 2013), and sometimes only poorly if at all related to theoretical concerns common in anthropological archaeology (Thompson et al. 2011).

Consider, for example, how routinely we use lithic and faunal analysis in anthropological archaeology. In those cases in which supervising anthropological archaeologists and CRM project managers are not themselves sufficiently adept at lithic or faunal analysis, they will consult outside professionals who are. One has performed both lithic and faunal analysis for hire for multiple third-party colleagues, and has found them to be very engaged, interested, and vocal in prompting input. Conyers (2012), however, relates very different experiences in relation to AGP consulting for anthropological collaborators: he describes failures to communicate, failures to engage in passing requisite information regarding site dynamics and properties of sediment matrices, and numerous other cognitive disjunctions between colleagues and AGP technologies. One has also encountered strange misperceptions about AGP among many AAs. One has noted that many of the questions asked and observations made by AAs to be only remotely connected to actual matters of AGP salience. Undoubtedly, the reason AAs often ask such tangential questions is due to unfamiliarity with the dynamics of actually conducting AGP surveys, the physical properties by which AGP technologies functions, the sorts of substratal conditions that can affect AGP technologies, sources of outside interference with AGP equipment, and hosts of other more minute considerations. Conyers (2012) clearly points to both AAs and GAs to become more attuned to one another's particulars in order to alleviate such miscommunications. One would reiterate that sentiment: to innovate current AGP methodology and to provide humanistic anthropological content to such technical applications we need both AAs and GAs to learn much more mutually about our respective research programs, theoretical frameworks, methodologies, assumptions, biases, and other cognitive factors.

CHAPTER 1

Geophysical Technology Commonly Applied in Archaeology

"I think you archaeologists need to get your noses out of the mud and stop collecting things."

—Dr. Balmurli Natrajan

FIRST: WHAT AGP IS EMPHATICALLY NOT IS A MAGIC BLACK BOX INTO which data are fed and from which mysteriously spring theoretically rich conclusions (Conyers 2004, 2006, 2012, 2013). AGP is a set of techniques and methods that can be applied to archaeological interpretations, to which we have to attune ourselves to gain explanatory power over archaeological phenomena in material and, increasingly, cultural contexts (Conyers 2004, 2006, 2010, 2012, 2013; Conyers and Leckebusch 2009). Significantly, AGP is much more common in Europe due to the completeness of sequence, length of record preserved, and technological familiarity (Johnson 2006). Less common, discontinuous, and somewhat infrequent American archaeological applications of AGP are related to general disciplinary unfamiliarity, confusion, expense, and interpretive difficulties (Conyers 2004). Common types of GP archaeologists use include geoelectricity, resistivity, EM conductivity, ground-penetrating radar, magnetometry, and magnetic susceptibility. Each method has particular selling points and advantages as well as inherent limitations (Conyers 2004; Conyers and Goodman 1997; Witten 2006). There is no one best method. What really varies is user experience, assumptions, and expectations, soil/sediment chemistry and structural conditions, and sampling increments of vertical depth and horizontal spacing. Sources of GP interference also vary. For EM methods, EM interference arrives from the sun and other cosmic phenomena, EM communications media, terrestrial radiation, and soil/sediment conductivity. For magnetic methods iron content of soil/sediment and nearby/buried metallic objects can certainly affect survey results. For seismic methods, vibrations from automotive traffic, heavy machinery, and tectonic activity affect results.

Geophysics has traditionally been used in several different ways in archaeology: Prospection for and surveying of archaeological sites; as a constituent of a suite of sampling and survey techniques applied to solving various problems at previously known sites; experimentally, to determine archaeological applicability and methodological refinement; and, increasingly, interpretively to assist in cultural resource management.[1] Given the destructive nature of excavations in archaeology, whereby it is often said that archaeologists are the anthropologists who kill their informants via destruction of material and spatial associations, destroying primary contexts, noninvasive AGP sampling techniques promise to become ever more relevant in archaeology. This means archaeologists should expect to be required to familiarize themselves with AGP at least to cursory levels of awareness in the future.

Most AGP applications have to date been conducted at well-known post-Neolithic sites with large-scale, monumental architecture, a clear case of picking low-hanging fruit (Thompson 2011, 2014a). There have been some Paleolithic open-air and cave applications published, but these are not the norm and they are few and far between (Conyers et al. 2013; Enloe et al. 2007; Thacker and Elwood 2002; Thompson 2011; Weinstein-Evron et al. 2002). There is also, due to peer-reviewed publication norms, a heavy emphasis on relatively *successful* AGP applications, stressing situations in which high-quality imagery of subsurface phenomena were produced, meaning that less successful applications or outright failures are largely unpublished (Conyers 2004 but see also Linington 1963 for an example of how long this bias has existed). On one level, this understandable bias is similar to the lack of anthropological archaeology publications regarding excavations in which no artifacts were found. AGP data, however don't actually produce *negative* results unless by faulty equipment or improper operation. The technology always produces imagery of something unless it's broken, and we should probably redefine success in multiple ways in regards to AGP. Both success and failure will be investigated below at a site in France and a site near Dubuque, Iowa. Yet, this measure of success as production of images or artifacts is problematic, because we rarely get a chance to learn *why* failure occurred.

A single book also cannot encapsulate the entire range of opinions archaeologists likely maintain about AGP, neither could it include all the things those interested in it would like to see happen in the field's future. Yet, in researching and writing this book, a deliberate effort was made to contact and collect the input from as many informants as was humanly possible (mea culpa, one only has so much time to send emails, and the law of diminishing returns prevented extensive follow-ups for failures to respond). What was asked? A very simple question was posed. Namely, one asked a variety of archaeological professionals 1) what they would personally want from AGP and

2) how would *they* improve it if they could, from GAs, to AAs, to CRM specialists, and as expected received a range of responses. It is extremely unfortunate that so few informants bothered to respond to the inquiry with which they were presented (well below 50%). There were probably a variety of reasons for why people failed to respond to the very brief set of questions they were sent, and since those cases involve null data no further time can be spent on them here. The identities of those who did respond will be kept strictly confidential.

Regarding one case in which a response was elicited, involving an Australian archaeologist, the professional apparently didn't actually understand AGP techniques nor did they see much utility in employing them or trying to improve them. In another instance, involving a mid-career American CRM professional, the response was that on one hand they didn't think or know if AGP needed to be or could be improved, and that on the other hand improving it "might put diggers out of business." One American AA expressed some frustration that one of their coauthored papers, which included a substantial AGP component, was poorly received by the GAs that reviewed it, and further criticized the idea that an anthropological article should subordinate anthropological theoretical content to geophysical purity (which seemed to be the reviewers' perspectives). Perhaps that AA should also question why the journal in question requested GAs to review an anthropological paper (do anthropologists routinely review geophysical papers?). Another American AA provided voluminous correspondence that covered a broad set of themes, including a fear, similar to thoughts expressed above regarding peer review, that their own geophysical submissions would never get published by GA purist reviewers, and that they were perhaps laboring in vain with attempts to publish AGP research. One lifelong American CRM professional intoned that all they really wanted or expected of AGP was that it "shows where the hot-spots are." Since Conyers (2010) and Conyers and Leckesbusch (2011) have provided strong arguments regarding potential for future anthropological geophysics, one accepts those sources and others (i.e., Kvamme 2008, Thompson et al. 2011) as a generally coherent and orthodox body of thought regarding how GAs view AGP. If that is a correct assumption, then it probably suggests that some GA professionals have nonanthropological biases, and AAs have anthropological ones. We need therefore to establish what would constitute a middle ground on which to meet.

It is important that archaeological geophysics develop and maintain a conscious anthropological footing. If AGP is not ultimately premised upon anthropological study of the archaeological record, then is it actually archaeology? Binford (1962) suggested that archaeology must be studied as anthropology. Does Binford still apply today? How does survey relate to anthropology? AGP has been certainly adopted as a preliminary survey/sampling method in American archaeology (Conyers 2004), during a familiar process of multidisciplinary "borrowing" of technology, method, and assumptions by which American archaeologists often come fill to their particular research programs. With respect to AGP, we should also note that this suite of technologies was developed for uses far outside the normal archaeological purview. In many cases, the technologies at issue originated in subsurface exploration for mineral and/or hydrocarbon deposits or other such "remote-sensing" applications (Conyers 2004). In AGP, most cultural deposits are encountered within two to three meters of the Earth's surface, as opposed to thousands of meters in depth as is often the case with petroleum exploration (Conyers 2004; Thompson 2011). This difference in analytical scales is also accompanied by vast differences in theoretical premises. Petroleum geophysics generally uses acoustic and seismic survey with very high energy, often explosive detonations for impulse generation and wave propagation through thousands of meters of Earth's crust to gather structural geological data of the subsurface lithostratigraphy (Mussett and Khan 2005). AGP is generally used to make sensible primary data maps of artifacts, features, and entire archaeological sites and landscapes that are very near the surface and that may not conform structurally to particular lithostratigraphic units (Conyers 2004; Kvamme 2003; Thompson 2011).

AGP is also not necessarily a "new" application within American archaeology, with the earliest electrical and magnetic subsurface survey attempts made during the 1920s and 1930s in Europe and North America, and continuing through the 1950s, 1960s, 1970s, and it experienced exponential improvements during the 1980s due to evolving computer technology and qualitative improvements in the reliability and capacity of digital storage media (Linington 1963; Conyers 2013). The long history of AGP applications in American archaeology is rather obscure, however, and not widely known, perhaps mainly due to the inaccuracy of early techniques, high expense, relatively poor methodologies employed, and restricted, unrepresentative "maps" and views produced by such early forays (Conyers 2013). AGP's early novelty also perhaps impacted its low notoriety in the United States. American archaeologists have therefore had to learn how to "do" AGP over time, and it is still very much a field in progress.

Yet, there also appears to be some consternation among some members of the traditional archaeological community (Conyers 2012), perceptible anecdotally but also in some instances in publication, that "geophysics doesn't work well for archaeology," or "Geophysical method X doesn't work for archaeology" (Collins and Molyneaux 2003). This theme that "geophysics doesn't work well for archaeology," or "Geophysical method X doesn't work for archaeology" is somewhat difficult to understand. One suspects that what this really means is that Archaeologist A tried AGP Method B at Site C and didn't get what they expected or wanted. Even where AGP is used "successfully" at a given site (and what defines AGP "success" is of course relative to the method and site in question as well as user *expectations*), the literature and anecdotal evidence (Conyers 2012, 2013, personal communication 2014) often reveals multiple archaeologists complaining that AGP didn't do what they *wanted* or what they *expected* it to do for them. Such examples are directly attributable to ignorance about archaeological AGP applications. This is but yet another index that AGP is still very new with relatively few practitioners in the United States, often with few lines of communication between parties. AGP and American archaeology in general (especially cultural resource management) would be served greatly by increased communication between AGP users and consumers. Enhancing lines of communication and increasing general awareness of AGP will be treated more fully in subsequent chapters.

In terms of expressing one's "ideal" vision of anthropological AGP, first one would suggest that they would like to see an AA and a GA in which transdisciplinarity was more the model of interaction as opposed to talking past one another (what Conyers 2012 suggests is the current norm). One would like to see all parties interested in AGP asking new and better questions, instead of "Will AGP work here or there?" or "What is the best AGP method for Site X?" One would like to see us asking and answering *humanistic* questions of the technology and our data set as well, such as "What can we learn about people and human behavior via AGP?" "How can we draw behavioral inferences about people from AGP models of subsurface material?" It is necessary that we ask such questions of ourselves. How can anthropological archaeologists inform themselves to be of greater benefit to geophysical archaeology colleagues? How can AAs contribute to increased and improved dialogue with GAs and contribute more to AGP applications? How can we more adequately communicate what we need from our GA colleagues? For example, Table 1.1 provides a possible outline of the sorts of questions AAs will need to ask and to answer to fulfill the goals of this volume.

As the reader will unquestionably note, many of the questions posed above in Table 1.1 require considerable a priori information and effort to answer, nor do they seem "anthropological" since they are based mainly upon material phenomena and AGP technology. One reason for this is that various physical conditioning agents have to be accounted for or negated before we move to attempts to assess human behavioral conditioning via AGP. Many of the questions also require multiple, sequential answers because they are complex questions regardless of their seeming simplicity. One should also note that this relatively arbitrary list of 10 questions can easily be expanded or contracted based upon actual field conditions at a site. For

Table 1.1. A Possible Outline for Asking Humanistic Questions in AGP Research.

1.	What surficial material indices exist at Site X?
2.	What do we now know about Site X prior to AGP survey?
3.	What do we *need* to know about Site X prior to AGP survey?
4.	What do we *want* to know about Site X prior to AGP survey?
5.	What *can* we know about Site X given its as-is material/spatial integrity?
5.a.	What do we know about Site X formation processes?
5.b.	What do we know about Site X post-depositional processes?
5.c.	What do we know about organic or microfaunal preservation at Site X?
6.	What graphic material indices does AGP survey elicit at Site X?
7.	Do features indicated at Site X by AGP vary spatially? If so, how?
8.	Do features indicated at Site X by AGP vary configurationally? If so, how?
9.	Do features indicated at Site X by AGP vary orientationally? If so, how?
10.	Do features indicates at Site X by AGP vary distributionally? If so, how?

example, many unasked questions inhabit the structures of what is proposed above, as in "What do we now know about Site X prior to AGP survey?" That question alone consists of many parts, such as: Do we have any information regarding the depth of the overburden emplaced upon archaeological materials? How does Site X relate if at all to other sites in the region? Do we yet have any information regarding site function? How big is the site? This outline is proposed only as a possible framework to elicit further thinking in an archaeological context, and to prompt researchers to learn as much as possible *prior to using AGP at a site*. If we as AAs intend to conduct our own AGP survey, or hire one done, we need to maximize what we know about a given site *before* AGP is employed. Our own subsequent experiences and that of our GA colleagues will therefore depend largely upon what we can know and relate to one another mutually a priori.

If we recall the suggestion from the anthropologist above that opened this chapter, one suspects that its meaning is more obscure and deeper than might seem, and not desultory invective. After reflecting on that suggestion for about 14 years, one suspects the actual meaning was that archaeologists need to think more and to dig less, to preserve the archaeological record of our shared cultural resources and more efficiently to use our cognitive equipment. That more widespread adoption of AGP could preserve or at least help to preserve intact archaeological data sets by partially obviating destruction of sites through excavation is a much anticipated goal (see also Conyers 2006, 2012, 2013). In order to realize the promise AGP offers we must work toward the development of new ways of using AGP and thinking about them anthropologically. This volume isn't the end of the conversations it seeks to elicit; it is just one possible beginning. For example, the four following questions can help to set a humanistic frame for AGP: 1) How and why do cultures interact with their environments? 2) How does cultural conditioning of the archaeological record affect the configuration, scale, orientation, distribution, frequency, and shapes of material features at archaeological sites? 3) How can we test such things geophysically? Finally, 4) How can we learn how to distinguish between anthropogenic patterning and natural, physical conditioning of archaeological material purely through geophysics? What questions would others ask?

Anthropology is a way to study people, and the ways in which it accomplishes its goals are at extreme variance with petroleum prospection or geophysical analysis of massive monumental structures. Anthropology is also mainly *about* people, as individuals, subgroups, societies, nations, and a species, but also how we think about people, study their complex behaviors, and try come to grips with ourselves. While GAs and their techniques obviously represent tremendous importance to our discipline, unless they mutually engage also in anthropological dialogue accessible to AAs in the context of their published research we have a fertile potential source for a communication gap. Likewise, when AAs fail to learn about geophysics either because they don't think they'll be useful, or they will somehow threaten dirt archaeology, or because they're "too expensive" (as if excavations are cheap), or for whatever reasons, we have other opportunities for absences of communication.

The communicative breaches that one describes are a result of many factors, occurrences with multiple causes. In many instances, geophysical archaeologists don't actually study people or even archaeology, inasmuch as they devote primary attention and literary primacy to demonstrating mastery of theoretical principles of wave propagation, electrical conductivity, and other physical phenomena. Unfortunately, many anthropological archaeologists are often quite ignorant of such specific matters and rarely if ever publish about them. The communication gap is also therefore a *literary gap*, in which subdisciplines publish research within rather narrow parameters that emphasize particularized paradigmatic, conceptual, methodological, and publishing characteristics that often fail to address the common problems faced by both anthropological and geophysical archaeologists in general. Regardless of our individual or mutual research programs and the time periods in which we work, we all face a set of common spatial and structural problems. For example, all archaeologists need to know the spatial dimensions of their project areas: how big is the site, what are its lateral and vertical extents, how does it fit within the surrounding landscape? We would likely all wish to know of how many related structural components is the site composed or is it confined to a single component? Given that we all face such problems we can all contribute to a mutual means for mitigating them.

Another contributor to the communication and literature gaps is epistemology. Anthropologists (including anthropological archaeologists) appear to approach the world and their research with a dizzying and often *changing* array of epistemologies, ranging from romanticism and idealism, to realism, and including pragmatism, Marxism, and in some cases nihilism (Trigger 1998). In some cases, AAs use multiple epistemologies simultaneously, the very fluidity of philosophical positions utilized reflecting the complexity of their human subjects. Geophysics, however, is much more a natural science, and as such employs a positivist epistemology, premised upon the field's "pure" theoretical pursuits and as an appendage of geology (Mussett and Khan 2000). The positivist epistemology has had effects upon archaeological applications of geophysics, in which problems are posed in primarily physical and structural terms as opposed to anthropologically or humanistically.

Anthropology is studied humanistically, and that includes tangible *material* aspects of being human (material, material effects, and material signatures) that we affect in our material paraphernalia, inherent to our condition as biocultural entities. Humans simultaneously inhabit temporal, physical, ecological, environmental, and cultural niches. Sometimes it seems to one that GAs, however, have focused much more on developing methodological applications specific to their own gephysical interests far more than human culture. If anthropology is the parent discipline of AGP, then perhaps we could find improved and more widely accessible ways to fit such research within anthropology, including research frames with humanistic, cultural foci as much as technical physical ones. One doesn't mean to be overly critical, but the primary scholarly journals for AGP can on cursory read seem to be written only for geophysicists, structural geologists, and mineral or petroleum speculators as opposed to anthropologists. This literary norm complicates communication between AAs and GA, thwarts wider accessibility in anthropology, and may even scare some anthropological professionals away from AGP. On many occasions one has heard anecdotes to this effect: AGP and its literature are "overly technical," although this is impossible to quantify. This structural bias in the AGP literature also makes it difficult for anthropologists to use the AGP primary sources unless they take considerable time and effort to acquaint themselves with them.

One does not, however, intend to place the onus only upon geophysicists to adapt to anthropology, for indeed many anthropological archaeologists could unquestionably and profitably acquaint themselves with the AGP scholarly literature and the equipment to the betterment of all interested parties. One admits the AGP primary literature corpus is highly technical and requires (at least of AAs) a substantial amount of appurtenant research to understand primary concepts, engineering, and the physical and material phenomena that

Table 1.2. Suggested conceptual foci to improve dialogue.

Anthropological Archaeologists	*Geophysical Archaeologists*
Basic physics and mechanics	Humanism
Dynamics of wave propagation	Environmental archaeology
Wave reflection	Site-formation processes
Electromagnetism	Taphonomy
Geochemistry	Functionalism/structuralism
Geology and earth structure	Evolution and neoevolutionism
Lithology and mineralogy	Processual archaeology
Geomorphology	Postprocessual archaeology
Geophysical methods	Social theory
AGP methods used in archaeology	Culture and cultures
Sources of AGP interference	Epistemology

underlie AGP. One doesn't think this potential dialogue will be as beneficial as it could be unless geophysical archaeologists and anthropological archaeologists mutually develop method and theory for addressing *common* mutual problems. In some ways, perhaps we need to develop a new discourse for communicating with one another. This can and should occur via what Conyers (2012) terms collaborative endeavors, in which teams of interdisciplinary scholars contribute reciprocally to their common problems. Table 1.2 proposes some of the mutual considerations and concepts on which I think we could mutually focus to improve dialogue between AAs and GAs.

What do anthropological archaeologists actually need from AGP? Well, one doesn't speak for all but they don't think we necessarily need in-depth knowledge of the inner operations of elaborate geophysical systems, although we do need to be aware of basics of such technology and to know general physical principles. AAs definitely need to know where and what kinds of buried archaeological materials can be expected at sites, roughly how large they would appear to be, and a mechanism for distinguishing between legiti mate anthropogenic deposits and various intrusive disturbances, such as prior excavations and landscape modifications. Most archaeologists don't actually need to know the theoretical specifics of pure geophysical theory (Mussett and Khan 2000), whereas they should know the basics; they certainly need to know where not to dig and avoid burning precious funds. Furthermore, since buried archaeological deposits can exhibit a wide variety of features in all manners of scales and configurations, we need to know how buried things look underground *across a very wide range of variation*. Walls, for example, are generally starkly obvious, linear features in AGP maps, as in figure 1.1, whereas lithic scatters and dispersed skeletal elements can assume numerous configurations, orientations, depths, and artifact frequencies (Conyers 2012). Walls, however, feature unduly disproportionately in the contemporary AGP literature. So one thing we need presently is a "comparative collection" of simulated and real archaeological features at small scales over a wide range of nonlinear configurational variations for comparing and contrasting material signatures. This would position such applications as middle-range research. This is most similar to how zooarchaeologists and lithic analysts use faunal comparative collections in their zooarchaeological analyses and lithic collections in lithic analysis.

One can readily find accessible descriptions of archaeological geophysics (i.e., near-surface geophysical technologies and methodologies applied to solving archaeological problems, see Conyers 2004; Gaffney and Gater 2003; Witten 2006). Recently, however, several challenges were posed to the AGP "community" (that body of archaeologists who routinely use geophysics in their research) in regards to becoming anthropologically relevant, with mixed or minimal results to date (Conyers 2010; Conyers and Leckebusch 2011; Thompson 2011; Thompson and Storey n.d.). There are in fact few research frames for archaeological geophysics presently pertaining to broader anthropological or social science theoretical perspectives (Thompson et al. 2011). We must recall that the application of geophysical technologies and methods was developed for other disciplines and its fit into archaeological research, although methodologically extremely powerful, is not theoretically seamless. Nor do all geophysics methods "work" for archaeology. Of those that do work in archaeological applications, often termed near-surface techniques, (Conyers 2006), one should sensibly ask whether the application has broader anthropological or humanist significance. Many don't. There are multiple factors in the mix. The point one wishes to hone is to find ways to do it more in terms of traditional humanistic anthropology *along with* new developments in AGP technique and sophistication for the betterment of archaeology in general. One would also assert that anthropological and geophysical archaeology should be *complementary* interactive pursuits, not merely parallel, non-overlapping ones so as to be maximally beneficial for all interested parties.

How should we set about finding theoretical and methodological means for 1) situating AGP inside anthropological archaeology, and 2) convincing AAs and GAs to forge a more productive mutual rapprochement? One is not writing this book as a geophysicist: first, because they are not one and second, because there are plenty such books out there that detail specific various methods for using geophysics in archaeology. We lack theory and even method for doing this *humanistically*, however, and one is an archaeologist who wants to use AGP in humanistic research to find out about people and human behavior without destructively excavating everything. Our goal should be more fully to integrate AGP technology into basic archaeological research design, field techniques, and even our own anthropological thinking, and new ideas for what we might do with it according to studying human behavior and people. Finding new ways to study the material inventory fruitfully to analyze archaeological signatures of human behavior, in middle-range fashion, is another goal. Of course, in making a statement like that one is in effect claiming that archaeology lacks such a book, or at least many such books. Otherwise, why would another be necessary?

Whereas one thinks many good books have been written on the use of AGP in archaeology (including several exemplary ones, such as Conyers 2012, 2013 and Gaffney and Gater 2003), few attempts have been made to provide theoretical or methodological framing for the use of such technology according to broader anthropological or humanistic goals. Anthropology and humanism, one submits, can provide valuable research perspectives for addressing some issues previous AGP publications have largely not. Combining AGP with humanistic anthropological perspectives should not require new anthropological or geophysical archaeologists to start from scratch, and this book seeks to provide some means to do both at once. This should be a cumulative endeavor. AGP can be relevant simultaneously to analyses of human behavior along with specifics of AGP technical phenomena, including human spatial organization, and more generally AGP "visibility" of material human residues. A biased question that one could pose would be what does AGP even mean if it isn't used to try and see aspects of past human behavior as opposed to presence/absence of material? Obviously, what we might "see" via AGP is conditioned and colored by any number of interactive physico-chemical, hydrological, and structural factors, but also by our own expectations, preconceptions, assumptions, and variable awareness of and relative expertise with AGP techniques. In other words, humanistic, anthropological, and geophysical methods of *thinking* are simultaneously equally as relevant as deep focus on and thick description of physical and material phenomena.

Per Conyers (2004:133), we must always ask ourselves—as both producers and consumers of AGP data—and answer whether or not the things we see in AGP data are germane to the research questions that compel our presence in the field in the first place. That is a tall order, and one that we may occasionally fail to accomplish. One suggests that what Conyers (2013) is suggesting is not so much a totality of a priori knowledge but rather an example of how more traditional archaeologists can *help themselves by putting themselves in positions to be more helpful both to themselves and to geophysical colleagues.* In other words, as much as humanly possible, we need to learn as much as is feasible about any given on-site conditions that can affect an AGP survey as soon as possible and relate them sensibly to surveyors before arranging such surveys. As has been noted above, failure on the part of supervisory personnel to determine site conditions (including soil/sediment conditions) prior to AGP surveys constitutes a major opportunity for blunders and communications failures and perhaps even resentments, especially when often steep price tags are involved. Furthermore, as Conyers (2012) relates, AGP technicians regard failures to communicate site and soil conditions as basic breaches of professional courtesy, in that money, time, and effort can be thoroughly wasted if even minor features such as kinds and levels of vegetation groundcover are not related. Perhaps other rationalizations exists for such failures by AAs, in that anthropological personnel tend to regard subsurface tests and excavations as *social* and *organizational/institutional* problems of manpower and money as opposed to chemical, structural, hydrological, or other fundamentally *physical* parameters that can adversely affect or completely preclude an AGP survey. For example, more help can generally be arranged for an excavation whereas it is not possible to remove the site sediment matrix and replace it with one more conducive to AGP applications. We all could do with some adjustment to our habits of thinking.

ONTOLOGY: ANTHROPOLOGICAL, ARCHAEOLOGICAL, AND GEOPHYSICAL

The establishment of anthropological reference frames for AGP involves interplay between anthropology (both as the parent discipline of archaeology as well as a legitimate social science in its own right), archaeology, and geophysics. Unfortunately, there are as yet not numerous common epistemic linkages between these fields. In my opinion, one reason that AGP has traditionally tended to lack anthropological content is due to ontological differences between anthropology, archaeology, and geophysics. To elucidate these differences fully and systematically would require numerous book-length works devoted to a wide variety of philosophical perspectives such as materialism, structuralism, functionalism, logical positivism, and empiricism in general, as well as some discussion of science philosophy. American anthropology, for example, at least sociocultural and linguistic subfields, is almost purely cognitive in nature. Its practitioners "work" almost exclusively with and within various forms of human cultural phenomena, both at individual and group perspectives, and tend to frame research goals in terms of culture theory and cultural context.

In anthropology, sociality and human group dynamics are emphasized research foci, for virtually all time periods. In regards to the relative ontology of Americanist archaeology, the Society for American Archaeology, one of the preeminent associations of American archaeologists, stresses in its principles of archaeological ethics a number of critical social and cultural criteria unique to its established materialist rationale.[2] American archaeology is devoted to analysis, explanation, and conservation of the archaeological record, and yet frames many or even most of its academic research pursuits anthropologically (Conyers and Leckebusch 2011; Thompson 2011; Trigger 1989). Regarding geophysical ontology, cognitive and/or social domains are largely ignored, since many of the applied aspects of geophysics are informed by strict scientific controls (i.e., rigorous hypothesis-testing and experimental replication) and are premised upon physical, structural, chemical, lithologic, geomorphic, and geologic premises, in other words, as a natural science. On the other hand, geophysics functions analogously to noninvasive diagnostic medicine, in which the observer uses remote sensing technology to image pathological somatic phenomena that are invisible to the naked eye but that often relate to individual behaviors (Witten 2006). AGP can, therefore, have an intrinsic cultural-behavioral component

A hierarchical assembly of theoretical levels can also serve as a useful heuristic. We might recall the hierarchical arrangement of anthropological and archaeological theories. There are high-level, mid-level, and low-level theoretical perspectives relevant to our discussion. In terms of high-level, what one might term anthropological, theory, would be a generalization such that diachronic human culture change or dynamics of human culture process can cause material culture changes and changes in the archaeological record. Low-level theory might be evinced by various hypotheses concerning the detection of evidence for material culture change archaeologically and through AGP methods, such as pattern recognition of behavior-mediated material signatures. In terms of middle-range theory, or epistemic linkage between high and low levels, we might suggest that 1) AGP can be used to identify some subsurface material cultural items and that 2) such material cultural items may evince perceptible change over time that 3) AGP might detect via changes in spatial patterning, material configuration and orientation, and scale (Thompson 2011:5).

ANTHROPOLOGICAL ARCHAEOLOGY: DIVERGENCE BETWEEN AMERICAN AND EUROPEAN PREHISTORIC MAGISTERIA

AGP is much more integrated in European archaeology than its Americanist counterpart. Perhaps one of the reasons European archaeology has earlier and more fully embraced AGP is due to the position of many European archaeology departments within the natural or biological sciences and greater familiarity with multidisciplinary research. In the United States, however, archaeology is normally situated within the social sciences, among sociology and anthropology departments. It is very interesting to note that whereas archaeology has long been essentially a middle-class prerogative in the social contexts of both Europe and the United States, the unique particularities characterizing historic trends in both places have led to somewhat divergent explicatory and paradigmatic magisteria (Binford 1962; Gibbon 1984; Trigger 1989). Given the very different historical contexts in which European and American archaeology developed, and, more important, considering the radically divergent archaeological sequences involved, we should expect numerous differences. In Europe, for example, the sequence of human occupation has recently yielded relatively early age estimates of nearly one million years before present (Oms et al. 2000). The sampling unavailability of large amounts of submerged European continental shelves in the northwest and along most of the Mediterranean periphery complicates interpretations of the European sequence immeasurably. In contrast, most evidence from the earliest North American occupation sequence suggest age estimates of only about the past 18 ky if not later (Fagan 2005). Consequent to this enormous temporal discrepancy between early human occupations of Europe and North America, the archaeological disciplines from both regions developed to solve very different problems.

Moreover, the social milieux of North America and Europe have been subject to very different and uneven historical processes which have had profound effects on the development of thinking within the respective brands of archaeology. Among European middle classes, the Enlightenment served as an immense influence upon the development of antiquarianism, geology, and archaeology, especially as it pertained to several key assumptions. European middle classes tended to view cultural patterns through filters of:

1. Psychic unity and basic social motivations
2. Cultural progress developing continuously over time
3. Progress in materiotechnological and sociality, economics, politics, morality, and even religion
4. Cultural progress imitates scientific progress through the elimination of ignorance, passion, and superstition
5. Reason as the proximate cause of progressivism (Trigger 1989:57–58).

In very important respects, North American developments in archaeological thought diverged considerably from the European norms above. Many of the earliest and most influential Euroamericans in the New World had themselves arrived in the Americas in the course of pursuing explicitly religiously themed pretexts, such as among the Puritans, Quakers, numerous Dutch religious sects, and other "pilgrims," whose journeys were at least nominally expressed in terms of a literal "exodus" from what were viewed as untenable (for them) social circumstances in Europe (Hildebrandt 2002). Many early European migrants to the Americas imagined themselves as "new Isrealites" and Native Americans as Philistines (Trigger 1989). The presence of Native Americans, as well as large-scale aboriginal material structures and even artifacts, was typically explained through appeal to a general Judeo-Christian interpretation as opposed to rational or scientific modes of explication.

It was only later in the Americas that New World prehistoric phenomena were sequentially subjected to rational and scientific interpretations, meaning that understandings of aboriginal American prehistory were perceived through explicitly colonial, religious, and lastly empirical lenses (Fagan 2005). The reduction of the largest chiefdom society in North America that was coeval with early European colonization (the Mississippian culture) through introduction of Old World diseases also obviated eyewitness accounts of such "complex" Native societies, to say nothing of earlier aboriginal examples (Diamond 1999). Since there were not living examples of Mississippian peoples witnessed at, say, Cahokia, all the easier for folk explanations and antiscience to conjecture all manners of allegedly Old World authors for Mississippian archaeological phenomena, as well as earlier large-scale earthworks from among Hopewell, Adena, Woodland, and earlier North American occupations. Hence the Mound-builder hypothesis (Fagan 2005), as well as other, mainly ideological, explanatory mechanisms invoked to account for the presence of Native Americans including a great many faked hoaxes intended to convince people that American aborigines had descended from ancient Israelites or even Egyptians. In any event, the prevailing thought was that Old World peoples had built the large-scale earthworks visible in North America.

It is pertinent also that European archaeology is traditionally ensconced within physical science or "natural history" departments that have long proceeded from explicit evolutionary systems of thought (Chazan 2011; McGuire 2008). American archaeology has instead tended to exist under the rubrics of anthropology programs at major universities. Cursory comparative reviews of graduate programs in archaeology at European and North American universities produce very different results in pedagogical design, while even undergraduate programs evince the same basic conclusions. American archaeology students generally learn the discipline within four-field anthropology departments taught as social sciences while Europeans frequently learn archaeology as natural science. The explicit focus on theoretical perspectives in American archaeology is also muted in the European discipline, which is largely focused on reconstructions of chronology and lifeways (Bogucki 1985). Many American anthropology departments have also recently begun reincorporating into even larger combination anthropology/sociology departments. How many American archaeology students learn as much about geology and geomorphology as they do about culture, language, cross-cultural awareness, ethnicity, politics (including intradepartmental feuds between social and material versions of anthropology), gender studies, religion, and other social phenomena?

Further causes for the lack of American AGP archaeological integration relates to cultural factors, including a longer history of use in Europe (Conyers 2013). Yet, this longer history of European AGP familiarity is also at least in part mediated by statutory obligations for AGP surveys that are sometimes actually mandated by law (Kvamme 2003). European archaeology students are also exposed to a much wider general availability of AGP equipment, databases, and use instruction at universities, libraries, and online (Kvamme 2003). Moreover, in Britain, due to the popularity and long duration of the television show "The Time Team" (broadcast 1994–2013), in which GAs perform AGP surveys at sites that are subsequently tested by field archaeologists, archaeologists and the lay public probably know more about AGP than most American AAs (Conyers and Gaffney 2000; Kvamme 2003).

One way we can seek to establish common reference frames is by analyzing cognitive elements common to anthropology, archaeology, and geophysics. They are not at all the same things. Anthropology, for example, is an enormous, holistic academic discipline that has set for itself the task of studying virtually all humans, alive, dead, and preborn (Nanda and Warms 2010). At least in the United States, archaeology is an aspect or a subfield of anthropology, premised upon recovery and analysis of the human material past or archaeological record (Trigger 1989). Many anthropologists and some archaeologists would, however, question whether archaeology really studies people per se, so there is not complete unanimity regarding that particular theoretical frame (Dunnell 1982). With regard to geophysics we should question what relevance we can demonstrate either to anthropology or archaeology. Can we find linkages that could bind anthropology, archaeology and geophysics? Such linkages will be seen to be extremely important to us because without them anthropology, archaeology, and geophysics exist as parallel disciplines instead of collaborative ones.

Anthropology, archaeology, and geophysics do share common ground (and the author hopes others will find even further examples of it), especially regarding *patterning*. All three disciplines study patterning and absences thereof. Basic to all three disciplines are attempts to define and recognize sensible patterning and violations of patterning (anomalies). There are differences in the respective pattern structures in the sense that geophysics seeks to detect patterning and anomalies in the subsurface based upon material, structural, and chemical properties (Musset and Khan 2000). Anthropology studies patterning in temporal, social, cognitive, spiritual, political, economic, and many other cultural domains. Archaeologists analyze many forms of material patterning, many of which have obviously been conditioned and impacted by sociocultural factors.

WHY A FOCUS ON SMALL SCALES IS IMPORTANT

Although the topics one introduces here will be expanded below, several key concepts guide the actual research presented. The author's bias is toward small-scale ephemeral materials. This is interesting and important and an underutilized component of geophysical archaeology. Most routine human behavior happens at small scales. Most humans do not individually as isolates or even in small groups mutually operate at scales of monumental architecture or entire communities. Each of us inhabits small individual and group spaces within much larger contexts of communities, societies, and landscapes. In order to learn about most human behavior and its material effects we often need therefore to analyze small spatial scales in our research. It is only after isolating and understanding material signatures at smaller scales that we are able to incorporate those into awareness of behavioral processes operating at larger scales, across wider spaces. As we will explore, however, much of the AGP literature is devoted to spatial analyses at very large scales, including pyramids, castles, bridges, and fortresses. Given the bias toward large-scale archaeological material in much of the contemporary AGP literature a very productive and crucial body of research remains for experimental and actual AGP research at smaller scales. Small-scale AGP therefore presents a very fruitful niche for exploitation by those who deal with ephemeral materials and features.

Foxhall (2000) presents another reason to focus on small scales: the theoretical tension between short-term human life and long-term archaeological thinking. We can imagine the archaeological record as a palimpsest formed by innumerable, short-term episodes, both isolated and recurrent, that developed cumulatively. If most of the behaviors about which we wish to reduce our ignorance occurred within small spaces during short intervals, then we need to think about and analyze smaller scales in order to gain better understanding. We should also be cognizant of the multiple, concentric short-term periodicities that we inhabit, such as seconds, minutes, hours, days, weeks, and months and the influences they exert upon our own material behaviors. Obviously, there are long-term, even lifetime, periodicities and "events" that we experience, but most are short term and occur within small spaces.

There are also important implications for small-scale AGP applications that we can derive from analyses by our colleagues in cultural and linguistic anthropology: namely, that human cultural, linguistic, and biological diversity seems to *decrease* over time. As a consequence, we should infer that human phenomena we observe now are truncated in comparison to much wider probable ranges of human diversity in the past. Therefore, if diversity *decreases* as we investigate intervals approaching the contemporary, then human diversity likely *increases* partially as a function of how deeply into the human past we look. Site scale also seems to decrease deeper in the past, with more recent sites featuring larger scales. A greater range of human diversity among sparse, widely dispersed, past human populations (including "archaic" humans) suggests a wide diversity of cultural/behavioral activities and wider ranges of material signatures to consider at smaller scales. Paleolithic open-air sites, for example, may feature wider ranges of small-scale variability than we have traditionally considered. Unless we "know" the total morphological, spatial, and structural diversity sites, features, and small-scale material signatures exhibit, how can we expect to identify or define such phenomena in AGP imagery unless we focus on small scales and *simulate* them and compile the images?

WHAT CAN AGP "TELL" US?

To answer the question posed just above, we need to think also about what we *want* to know, what we *need* to know, what we *can* know, and consider those against what AGP can show us. In order to frame our discussion, however, several key points that shall be further developed below now need to be presented. First, AGP cannot actually "tell" us anything. Geophysics do not provide underground pictures and they don't provide instructions for what to do; what these techniques provide are *models* of the subsurface, models that we need to use along with other tools in order to get the most out of them. We also have to extract ancillary information from our own knowledge and relative experience in our areas of expertise, from published examples of other AGP applications in the professional literature, from published content from the AA literature, from our own "traditional" maps, including GIS, hand-drawn plans and stratigraphies, photographs, pedestrian surveys, shovel tests, augur tests, and technical analyses of our material inventories, and combine all of these into a coherent "picture" of our sites.

A number of excellent sources do exist to assist us so we don't have to reinvent that many wheels. In an excellent (if dated) overview of geophysical technology for scholars and laypersons (including archaeologists), Mussett and Khan (2000) suggest that potentials for incorporating geophysics in archaeology relate to the *modeling* of the likely presence and dimensions of archaeological sites. Such capability also includes the opportunity to map such criteria, and, perhaps even more important in excavation contexts, showing which areas of potential sites may be most productive in terms of labor and capital investments. Yet, such techniques can also be applied productively at previously known sites where excavations have already proceeded. There exists an obvious importance here for CRM, because under traditional survey/sampling techniques important material can be and often is completely missed. Conyers (2006) and Kvamme (2003) also advocate the focusing of these powerful technologies for knowing in which areas to *avoid* testing (perhaps just as important as or even more so than models of where to test), and as a method for inserting sites into larger landscapes. The latter is crucial because it involves moving from local analyses of sites as "point estimates" of specific isolated variability into the establishment for how sites fit into larger, regional spatial contexts, while the former is the bane of practically every archaeologist. Mechanisms for showing us where *not* to dig are of inestimable value, especially for CRM professionals who work under tight fiscal and time constraints (not that we all don't).

Time is just as much a focus for archaeologists as are geophysics, site reports, method and theory, paradigms, and the litany of various "-isms" to which we subscribe. Such as it is, we have to deal with concepts of time. We can't know the future (at least in most of our lifetimes), unfortunately, because it hasn't happened yet and effects thankfully don't normally precede causes. We can't really analyze the present (despite admonitions received via the pop media to live within it) because the present is a flux medium, existing as a state of perpetual flux between the future and the past states of being. The past is what we can actually analyze, for analysis takes time, and the past is temporally forgivably static. But the record of the material past changes . . . it's not static.

So what is the material-archaeological past? Materially, it's just a scattered, compressed, eroded series of things associated in proximity, only partially indicative of events that transpired in particular places at particular times. What does this mean for AGP and people who might try to use it? Conyers (2013:167) suggests that the pressure is on *us*: to understand AGP data acquisition, signal-processing, and methods of interpretation and then to be competent to explain such things to other interested parties. To be explanatory or really useful, AGP techniques have to be integrated and embedded within other anthropological archaeology research frames (Conyers 2004; Thompson et al. 2011), including humanistic anthropological ones. That's another of the volume's goals: to help pave the way to greater integration of AGP into traditional archaeology, to obviate as much as possible digging and informant-killing, so that future archaeologists who will possess new theoretical, methodological, and technical means can process the archaeological record on behalf of all.

One will argue that humanistic archaeology can make use of geophysics to study subsurface material patterns, and that in some cases we can know that such patterning is illustrative of different human behaviors that are distinguishable from natural physical taphonomic forces, enacted over the short term at small scales. Such an interpretation is inherently anthropological, even if the material patterns we analyze and the putative human behaviors that partially conditioned them occurred in the past. How can we do this, one might ask? Well, archaeological geophysics happens to be a very good way of "seeing" change in expressions of material culture though time. We can see changes in how landscapes or parts of them were used via AGP. At some sites, given good spatial integrity, we can investigate how the scale, configuration, and organization of buried material vary across space and depth, and when combined with data from excavation and technical analyses (faunal analysis, lithic analysis, etc.) we might be in a position to speak to differential material expressions as functions of differing site occupational scheduling and seasonality, different periodicity, or other factors. Such data and interpretations can also be cataloged and archived permanently for future use by other archaeologists, geophysicists, geographers, sociologists, and specialists, and this is also inherently anthropological and interdisciplinary.

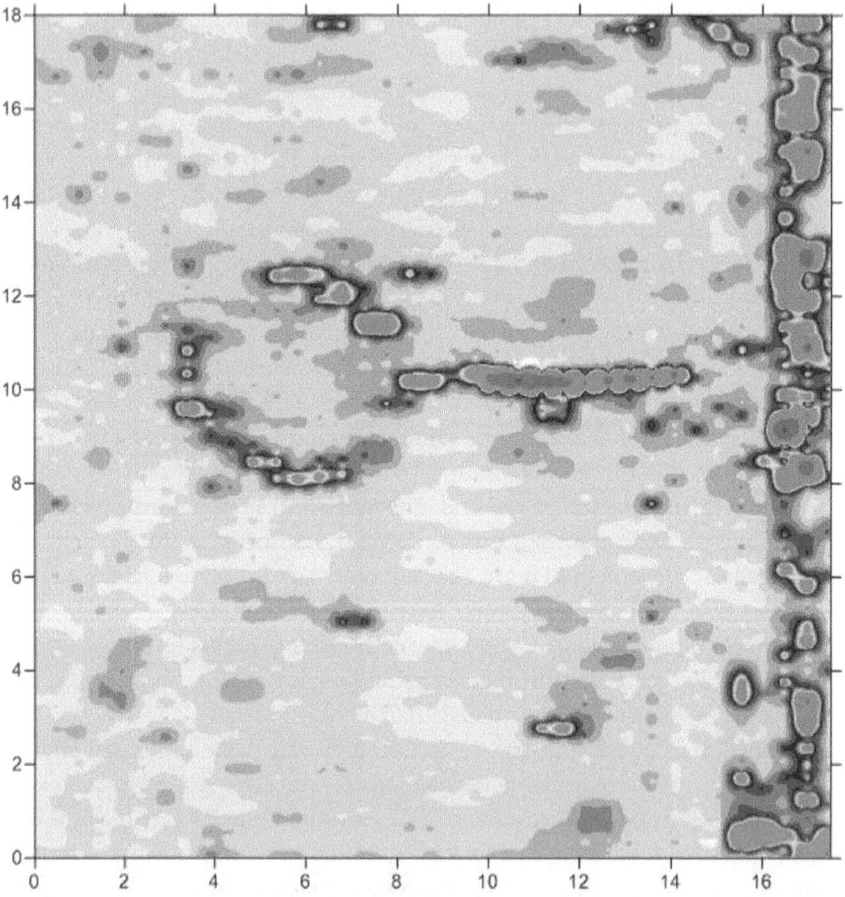

Figure 1.1. 400 MHz radar amplitude-slice at 9–10 ns time-depth 2 ns in thickness, collected at the Gangivecchio monastery site, Gangivecchio, Sicily. Note the obvious wall at right, and the piscina or fish-pool at upper left-center and the paved walkway at upper center. Walls are frequently not subtle features in AGP data.
Data generously provided by Dr. Glenn Storey of the University of Iowa.

NOTES

1. Abbas et al. 2005; Bevan and Roosevelt 2003; Boniger and Tronicke 2010; Carrozzo et al. 2003; Chadwick and Madsen 2000; Chapman et al. 2009; Chianese et al. 2004; Conyers 1995, 2004, 2006; Conyers and Cameron 1998; Dogan and Papamarinopoulos 2003, 2006; Due Trier et al. 2009; Forte and Pipan 2008; Gaffney et al. 2004; Goodman et al. 2006; Grealy 2006; Hildebreand et al. 2002; Hill 2001; Kvamme and Ahler 2007; Leckebusch 2003, 2005; Leckebusch and Peikert 2001; Leucci 2006; Lorenzo and Arias 2005; Negri and Leucci 2006; Neubauer et al. 2002; Piro et al. 2000; Pomfret 2006; Rodrigues et al. 2009; Rowlands and Sarris 2007; Ruffell et al. 2004; Storey and Whittaker 2004a, 2004b, 2005a, 2005b, 2005c; Thacker et al. 2002; Thompson in press; Thompson 2011; Vafidis et al. 2005; Weaver 2006; Yalciner et al. 2009; Zhou and Sato 2001.
2. (SAA website 2012http://saa.org/AbouttheSociety/PrinciplesofArchaeologicalEthics/tabid/203/Default.aspx)

CHAPTER 2

Bridging Gaps

In one's opinion, the GA and AA communities of professionals in the Americas need to bridge the gap between the more technically demanding, engineering- and physics-dominated applications of archaeological geophysics and the more humanistic aspects of anthropological archaeology. Both subdisciplines could be stronger for the effort. Although AGP is becoming more accessible to AAs, especially ground-penetrating radar largely through the work of Larry Conyers (2012), much if not most of the AGP literature is still highly technical in nature and only tangentially concerned with identifying possible human conditioning of subsurface archaeological materials, or even focused on people: instead, the onus is placed on the identification of subsurface archaeological material and the exhaustive description of various wave parameters and equipment arrangement and settings. There is, however, unquestionably a learning curve associated with the acquisition of AGP proficiency, both in regards to the technical demands of operating sophisticated equipment as well as the subjective interpretation of subsurface reflections off buried material and what are traditionally termed anomalies in magnetic and electrical methods.

Please allow one a bit of descriptive and digressive license here through appeal to a metaphor. The use of total-station theodolites is becoming increasingly common in the repertoire of archaeological methodology. This awesome technology adds speed and precision to excavation, especially where large quantities of material need to be excavated and processed rapidly. For example, at the Magdalenian site of Verberie in France, one has frequently witnessed many individual 1 m^2 units containing *thousands* of artifacts, generally to the frustration of the excavators (see Audouze and Enloe 1997 and Thompson 2011 for examples of the dataset sizes). Now, imagine if publications of archaeological datasets acquired with the assistance of total-station theodolites devoted most of their text to descriptions of the specific equipment settings and operating methods used to calibrate and manage the equipment as opposed to discussions of the actual materio-spatial data acquired through their use. We might add to this some basic descriptions of subsurface imagery as well as AGP image models of buried material. While somewhat satirical, the preceding does accurately portray much of the AGP literature, especially certain of the primary journals and noteworthy books, which are largely dominated by discussions of equipment settings arrangements, specifications, and pictures instead of *archaeological material* or the humans that partially arranged it. Then imagine that the people who wrote such articles and books seemed to direct their writing not to other archaeologists or anthropologists but intended it instead for engineers and geophysicists.

Conyers (2012), in particular, has described a general difficulty in explaining AGP techniques and even in facilitating basic communication of important topics with what he terms traditional or "other" archaeologists (clearly as opposed to geophysical archaeologists). The premise of this and similar miscommunications is clearly based on the general lack of AGP awareness on the part of archaeologists who are primarily *anthropologists* with humanistic research foci. So stipulated, Conyers (2012) explains a portion of the present communication gap in completely valid fashion without addressing some other considerations: have geophysical archaeologists or geophysicists thought out a basic series of precepts whereby geophysical data can be readily incorporated in archaeological *explanations*? Archaeology has largely concerned itself with explanations of and no longer simply documentation and description of past anthropogenic material phenomena. Who should be responsible for developing such thinking? Is that primarily a geophysical issue or an anthropological one, or does it have mutual import for both fields? If both camps need to weigh in, then how can we negotiate a process to build such a body of collaborative thinking in mutual fashion? It seems obvious that both GAs and AAs need to be together in the loop on this formulation.

How *should* anthropological archaeologists address geophysics and on what terms? What common ground do we share? Geologists and geophysicists have devoted years if not decades of training and apprenticeship to become conversant in the intricacies of their respective professions, as have anthropologists and archaeologists. We're all highly trained, what differs is the specific content of the training. Anthropologists, however, are not generally accustomed to dealing with geological intricacies or even largely with material phenomena at all. AAs have somewhat different sets of tools in their toolbox than anthropologists and a thoroughly specialized and distinct jargon all their own because they do deal with material phenomena. When geophysical archaeologists arrive and survey sites with their technologies, what specific anthropological-archaeological material expertise or cultural is required for them to do their job? Are they trained in archaeological techniques and have they participated in excavations? Do they know the differences between artifacts and features, flakes, blades, and cores? Such questions are posed honestly, for while AAs often betray some ignorance of AGP specifics, how many GAs have trained in the specifics of anthropological archaeology? AAs are often unfamiliar with geology, lithology, geomorphology and surface processes. And they shouldn't be. Few archaeologists, indeed, and even fewer anthropologists worry over (or have conditioned themselves to understand) how various structural geological constraints and geophysical equipment settings and parameters that directly affect the efficacy of the personalities and technologies they (usually) hire to come post facto to their sites and survey the subsurface. One seeks therefore to bring both camps closer together in service of both anthropological and geophysical archaeology, in collaborative fashion, for production of results that could not be obtained by either subdiscipline alone (Conyers 2012; Isayev 2006). One of the ways of overcoming differences is analysis of them, so we can mutually participate in integrating anthropological and geophysical archaeology.

To the extent that anthropological archaeology has been defined, Gibbon (1984) provides some very relevant aspects of a definition, including: adoption and refinement of the culture concept; syntheses of cultural ecology, structural functionalism, and cultural materialism; new assumptions about the role of archaeology and subject matter; new conceptions of the scientific method; increased borrowing of methods, theories, and techniques from other disciplines. In practical terms, the anthropological questions we ask in guiding our research designs establish the premises by which we attempt to provide explanations. One goal of an anthropological archaeology would be to move beyond material data recovery, seriation, and description to understandings of human behavior that conditioned such materials. In this way, anthropological archaeology is partially commensurate with Binford's (1983) Middle-range research, in which we infer former human behaviors and activities from archaeological material signatures. This is something we sorely need in AGP, and this will be developed much more fully below. We have to ask anthropological questions in AGP research. Problem-solving and prospection are laudable goals but so too would be a greater focus on human behavior as opposed to descriptions of material (including the equipment) or method.

Does AGP have anthropological relevance or standing? Conyers (2012, 2013) indicates that it does, or at least that it should. But who decides if or when AGP has an appropriate anthropological relevance or standing? *Both* anthropological archaeologists and geophysical archaeologists have to contribute to a new collaborative dialogue. Joining that dialogue requires some discussion and negotiation, but mostly it comes down to basic goals and assumptions. If most of the communication between the two sets is nil, then how would anyone be in any position to say? This is a problem, and it may only be alleviated through increased communication and awareness between AAs and GAs. I think a noteworthy aspect of this dilemma is related to the nature of most AGP studies heretofore performed, which is to say they have been mainly dedicated to "problem solving." If one is studying a partially exposed wall, for example, some unknown portion of which lies beneath the ground surface an AGP survey might tell me how much if it is still buried and solve a dimensional problem. An AGP survey might also give one some clues about surface or horizon continuities and discontinuities, or whether and how much erosion or tree roots have impacted the wall. Yet, what will such surveys tell me about the process by which the wall was built in the first place? Let's assume the wall was built in stages, perhaps even by different groups of people at different times, and that various material and structural differences are visible in the exposed portions of the wall. How much of the stylistic and/or material differences between building episodes will be revealed in the subsurface, if they exist? How could they be identified in subsurface AGP data and images? And how can one use AGP technology to get at any social dimensions of the humans who built it?

The bias toward a narrow concept of success has removed many opportunities for the discipline at large to learn from negative results, which can sometimes be just as informative as successes. While the majority of scholarly attention has to date been directed toward refining AGP subsurface imaging capabilities, and integrating AGP with other archaeological field methods, such as pedestrian survey and GIS applications, generally using very large-scale structures at very well-known sites (for example, on the Giza plateau of Egypt, or a wide variety of buried Greco-Roman structures such as walls, temple foundations, etc.), to the virtual exclusion of most older smaller scale structures and features. Geophysical archaeologists should be queried about the post-Neolithic, large-scale publication bias, but the bias itself is understandable. Consider how the development of what became anthropological archaeology as a discipline earlier took nearly the same path: differential use of a wide variety of techniques at primarily large-scale permanent sites, visible at the surface, such as at Knossos, or Troy, or upon the Giza plateau. It was only later that some scholars began concentrating on earlier times such as the Mesolithic and Paleolithic, lacking monumental architecture as a landscape guide to where to dig. Consider also how much of the archaeological literature consists of later scholars piling on the heads of the earlier ones for ill-formed research designs, poor methodologies, and unrepresentative sampling strategies. Consider how long it took for prehistorians to recognize Paleolithic stone tools for what they are!

In retrospect, the discipline of archaeology could ideally have learned much about itself and much more about the archaeological record if it had been undertaken as a development from the earliest Paleolithic to later periods, moving from analysis of small ephemeral sites to examining large-scale phenomena. Instead, the real situation is what we have. In hindsight, then, materials- and physical-properties-based geophysical archaeologists should be forgiven for some of the publication bias. Reviewing relevant early AGP literature indicates that this bias toward successful (=pretty and informative pictures) occurred almost immediately (Linington 1963; Kenyon 1977). It's nearly built into the fabric of AGP literature.

The AGP literature also indicates that the 1990s and 2000s saw a kind of "gold rush" at work, as there was a dash to stake various claims on particular field methods, and sampling techniques. for publications before much experimental AGP research was completed (Conyers 2010; Thompson 2011). Several of the most interesting experimental works published during that interval, however (Leckebusch and Peikert 2001; Grasmueck et al. 2004), suggested that AGP could operate at smaller scales, even at the level of individual items or small sedimentary structures, provided that adequate experimental controls were exerted on research designs. Subsequently, some attempts were made to publish truly experimental research (Conyers 2006, and see especially the methodologically exemplary Solla et al. 2012 forensics article as another example).

Unquestionably, the vast majority of the literature through that interval reflects "successful" imaging of buried materials but no real discussion of any experimental or methodological refinement outside of getting better resolution in imaging. It's as if the images were the *only* goals instead of ascertaining understandings of human behavior that conditioned the various configurations of subsurface material in the images. There had been very little discussion of anthropological goals for AGP survey to date, in fact few discussions of AGP uses *beyond* surveying and prospecting, until the late 2000s. In other words, it appears at least superficially once various methods for collecting raw AGP data in the field were developed they rapidly spread among users who appear to have virtually replicated the same basic methodology at other sites with large-scale materials. This methodology would appear to be, at least in general, something along the lines of Step 1) take various geophysical equipment to the field, 2) gather and process raw data into images, then 3) publish the images along with some basic explanatory text concerned with wave dynamics, with no real attempts made to connect the research to larger archaeological or anthropological theoretical issues. In one's opinion, AGP journals and publication preferences bear considerable blame for this, as do adversarial peer-review norms, as described by the informants above, in reference to caustic peer reviews that failed to take any account of anthropological merit. This bias can be confusing because "unsuccessful" publications, at least those that were not published, cannot be examined. Was an entire study unsuccessful or just part of it? Was the lack of imaging success due to an absence of subsurface archaeological material? Which AGP technique(s) were used? How would we know?

The publication bias toward perceptibly "successful" applications specific to AGP comes at a serious cost also in that it has perhaps caused an expectation among nongeophysicists that every AGP application should be successful, as thought there *were* no "failures." Nor were "success" or "failure" actually defined beyond the ability to image things buried in the ground. To some extent, this is true: there are no AGP failures unless one is using defective technology or there is some structural, chemical, or other interference that thwarts sampling, or even inadequate user training and data processing that prevents usable imaging of the subsurface. Unless one is willing to invest the requisite time, money, and effort into processing raw data into useful imagery the final results will be discouraging. We cannot compile a comprehensive database of what works as opposed to what doesn't unless we get more opportunities to learn from unsuccessful applications. Nor are we able to compile an exhaustive comparative dataset of AGP images that can be compared with those from other sites, which should be a primary focus of such research. We must understand that publishers are also under numerous constraints regarding what they publish, yet although it is understandable that "less sexy" AGP failures or even partial successes are withheld from publication it nonetheless contributes to the bias in the AGP literature.

In reality, the ultimate relative success will be the ways in which users demonstrate phenomena of anthropological relevance through the use of AGP techniques in archaeology. In the focus on AGP "success" vs. "failure" we often fail to analyze whether or how AGP is embedded within larger anthropological and archaeological paradigms and research. Are all instances of "unsuccessful" AGP applications otherwise devoid of all anthropological merit? How would we know if we don't read about them? Are "successful" AGP applications inherently anthropological? The reader can judge the difficulty in assessing AGP that is premised upon success as opposed to anthropological research premises. The invisibility of "unsuccessful" AGP research robs all of us of opportunities to learn from one another and to help each other. It is a serious problem and one not terribly conducive to easy solution. Under tight monetary constraints, journals are simply not going to publish material that is regarded with the taint of failure. We should, however, establish among ourselves a professional database of AGP images from the widest possible variety of sites and time periods to serve as an open-source comparative collection similar to faunal and lithic assemblages, including "failures."

Without completely tipping one's hand in this chapter, we have to turn the discussion from geophysics back to people. Anthropology studies people. Most of the research framing will have to do with the questions we ask of ourselves and our data as we formulate methodology. For example, how many things do people do that can be expected to leave traces visible to AGP at a later date? What things do we use, in what associations, patterns, and frequencies, might be detected later through AGP? Then, just what are the relationships between us, our various activities, and our things? How can we abstract these? How can we use AGP techniques in tandem with other established technical analyses to learn more about the archaeological record and the people who made it? What things can we *not* do with AGP?

As above, several GAs have explicitly challenged geophysical archaeology to answer anthropological questions (Conyers 2010; Conyers and Leckebusch 2010; Kvamme 2003; Thompson et al. 2011). Previously explicitly anthropological foci in AGP hadn't been a focus, but even granted that, what questions should we ask and what will this collaborative endeavor look like? One submits that questions of methodological integration can guide the development of such collaboration: How can archaeological and near-surface geophysical methods be used more frequently in the United States, complementarily along with pedestrian survey, aerial photography, and excavation to address issues of anthropological significance, such as social organization and site occupational scheduling (seasonality)? Related questions of crucial theoretical importance would also be: What sorts of human behaviors and events leave geophysically visible traces or material signatures in the archaeological record, how should we look for them, and what do they look like when imaged geophysically? There are no intrinsic reasons why archaeological geophysics cannot become an anthropologically and theoretically informative methodology in its own right, just as have pedestrian survey, zooarchaeology, geoarchaeology, and lithic analysis (Thompson 2011).

One is a traditionally trained anthropological archaeologist, who received teaching in both theoretical archaeology and CRM applications, as do most American archaeology students. So one has their own biases and certain perspectives that have been developed independently and yet some that have been more inherited from one's mentors. It is therefore significant to note at this point that one's mentors were also anthropological archaeologists. One would, however, ask our more "geophysical" AGP colleagues (i.e., those who typically do research that is published most often not in the mainstream archaeological journals but in the more technical geophysics and engineering ones) just how they view AGP operating inside archaeology. What do you want geophysical archaeology to be inside the larger field of American or Western archaeology? One would then ask our anthropological archaeologist colleagues and ask what it is that they would like to see from AGP. If it seems as if one has positioned themselves between two poles, they have. Intentionally. This was done did that so one could ask mutually just what it is would we like AGP to do in archaeology and how should we do it?"

CHAPTER 3

The Nature of a Problem

By way of comparison, any attempt to frame archaeological geophysics inside an anthropological research paradigm is in some ways analogous to using ultrasound or other medical tomography devices to assess a person's behavioral or sociocultural situation. One thinks such a program is possible within some sets of assumptions, however, whereas it might also be necessary first to postulate certain assumptions about the potential physical, somatic manifestations of various behaviors that medical tomography could detect. Medical tomography devices might detect the presence of foreign matter in the lungs, for example, or of lung tissue deterioration, due to smoking, say, or inhalation of airborne particulates from a coal-mining lifestyle. The technology can detect such somatic phenomena but it cannot impute causal significance to material detected. Without knowing whether a person's evidence of lung disease was due to a particular cause, we would need either to assume a cause, compare the detected physical manifestations to databases of other, similarly afflicted patients and infer causal similarity, or obtain some testimonial evidence regarding potential causes from the person themselves. First-person testimony would carry the most weight in this scenario, but people lie. Estimating causal similarity is basically the use of argumentative analogy, a process very familiar to archaeology and anthropology, and is especially similar in many respects to Binford's (1977, 1981) middle-range theory. Assumptions of potential causes are least reliable, because we can be simply and often are completely wrong in our initial assumptions.

The above presents us with a problem, familiar to archaeologists, in which we must use nondocumentary evidence to impute the proximate cultural or anthropogenic cause of various physical phenomena (Binford 1983; Schiffer 1976; Trigger 1989). Equifinality is obviously an issue as well, since we often can't actually be certain which anthropogenic process ultimately formed a given site or whether physical processes formed it. Realistically, however, this does not present that much of a problem. A useful way to look at it is that we must use linking arguments (middle-range theory) to connect observations (geophysical imaging) with our data (the geophysically imaged archaeological record). Assuming that a given site will be ground verified post hoc, ample opportunity exists for falsification of incorrect or unwarranted hypotheses. Since geophysical analysis is basically just another form of sampling, we could regard verification of results as a form of hypothesis testing based squarely upon the geophysical imaging of portions of the archaeological record (Binford 2001). As with all archaeology, AGP cannot pretend to study people. It produces subsurface images of portions of the archaeological record that was partially conditioned by people in the past. These images are in fact not pictures but models. This will be discussed in more depth below.

All buried archaeological material (including the excavated inventories), from the smallest lithic or bone artifact, to small-scale hearths and other domestic features, to the foundations of the largest monuments are all part of the archaeological record. Given that stipulation, we can ask how we study the archaeological record. This book uses Binford's (1977, 1980, 2001) and Schiffer's (1972, 1976, 1987) middle-range theory and research strategy, which is a program of inference and the formation of linkages between observations and theories, between uses of complementary data acquisition and analytical methods, and between the archaeological record and conclusions drawn about it. A very recent and otherwise extremely high-quality journal article (Solla et al. 2012) outlines a basically detailed program for conducting middle-range AGP research in a forensics context. Unfortunately, this otherwise exemplary technical and experimental article completely failed to frame its research in middle-range terms.

One considers Conyers (2010) and Conyers and Leckebusch (2010) as very serious anthropological challenges to some established AGP orthodoxies. We *must* deal with the anthropological framing issues of potential research very seriously if we expect to derive any realistic, practical professional use from it, and we should consider very carefully (as above) the sorts of questions we wish to ask of the technology. There is no rule book for any of this. There is as yet no anthropological conventional manner for incorporating geophysical studies into cultural resource management or theoretical archaeological applications; *we have to build it.* Anthropological geophysics is therefore ambiguous, as it is in the process of being constructed. Another cause for its ambiguity is the graphic, image-intensive nature of geophysical archaeology. It consists of pictures, of images, of graphically arranged portrayals of physical data that pertain uncertainly to the actual material in the ground, and perhaps even less to the sorts of interpretive problems anthropological archaeologists encounter and pose to themselves. For that matter, AGP images are not really pictures, they are essentially models (as we shall see below) of physical reality produced through many cascading human cognitive and computer software processes.

There could be no rule book to any of this. For AGP was approached piecemeal, on virtually a site-by-site basis, and the research involved was published with the peculiar success-driven paradigm (i.e., success being the production of accurate and sensible imagery) developed within the politics of academic publication. Were the AGP subdiscipline's theoretical bases generally analyzed by colleagues from other subdisciplines, such as anthropological archaeology and CRM? As regards CRM, moreover, where archaeological geophysics can possibly make its greatest practical application, it appears there exists little substantive communication between CRM project managers, executive directors, and principal investigators on the one hand, and geophysical technologists (including geophysical archaeologists) on the other outside of summarized reports. Why should this be the case? For the CRM professional, AGP offers perhaps one crucial opportunity above others: the ability to determine where *not* to dig or what sites to avoid excavating, for any numbers of reasons. This hits the bottom line directly, and we are certainly not discussing something tangential to archaeology, project design, and survey deadlines. At issue are basic understandings of the advantages geophysical archaeology offers, as well as mitigating, complicating factors unique to each particular CRM (and others) application. Understanding how geophysical archaeology "works" therefore involves better, more thorough, understandings of particular site contexts as well as AGP techniques themselves. For before one contracts a geophysical specialist, one should already have a clear view of actual site contexts and likewise a commensurate appraisal of the potential utility of any geophysical surveys that might be planned or conducted at a site. Such goals demand a thorough awareness of project areas, contexts, objectives, and physical parameters, but AGP offers convenient, practical, and rapid means to avoid burning CRM funds.

NOTHING TO EXCESS?

We have to deal with the fact that archaeology involves humanism and relativism as much as it does science: perhaps to excess, perhaps to an insufficient degree. We are unlikely to find ultimate resolution because we each have different ideas about how the discipline should operate. Archaeology has had indulgence in a lot of excess; that much is certain. Our discipline began as an antiquarian pursuit for absurdly wealthy people who had developed fascination and fetishes with the past, and especially Mediterranean and Near Eastern pottery. It then evolved into a documentarian and descriptive pursuit, whereby those with skills as documentarians and describers documented and described various attributes of numerous old things that were found either on or in the ground, especially from burials, temples, and other sacred spaces. Archaeology then entered a very valuable self-analytical phase, in which it asked itself whether it was science, anthropology, both, neither, or something else entirely (Trigger 1989, 1998). Carried along for that gut-wrenching ride were phalanges of graduate students, mainly at, increasingly, southwestern universities but also a few northern and eastern ones, steeped simultaneously in traditions of documentarianism, descriptivism, culture-history, and very astute instances of collegial self-reflection. Much of this disciplinary self-reflection had to do with the specific philosophical traditions, among many "pure philosophy" suitors, to which archaeology was trying to hold itself betrothed (Flannery 1982; Salmon 1975; Wylie 1985).

Then occurred the 1980s, when processual archaeology came to relative adulthood along with an explosion of new methodologies, when post-processual archaeology emerged in response, along with the emergence of corporate "CRM firm" archaeology. It was in this milieu that AGP became a little more frequent in research programs. Looking back at it that way, it is no wonder that many archaeologists have little bearing in regards to AGP. It might represent to many just another technique, just another methodology, whereby past data can be processed, and never mind the specifics of how, where, and why it can be informative. It was another "tool in the toolbox," so to speak, another lever to pull, another button to push, and, since, according to the relativists and deconstructionists, this isn't actually a science but a deconstructed "mental construct" who should be the wiser? Given the profusion of archaeological technical methods that developed during this period, anthropological professionals could be forgiven for overlooking AGP, at least to some extent. In the midst of zooarchaeology, bioarchaeology, spatial analysis, lithic analysis, refitting studies, use-wear analysis, replication studies and other ethnoarchaeology, and many other specific research avenues, perhaps AGP tends to get lost. Perhaps its time had not yet arrived, given the bulkiness and unreliability of many digital devices of the 1980s and early 1990s.

Anthropology, however also changed over this interval. How? One possible answer to this is framed by Crapo (2013), who elicits the structure of anthropology as being a dualistic amalgam of both science and humanism. Anthropology incorporates aspects of both perspectives into one ontological entity. It is also the biocultural study of humans, in all places and times (Park 2008). We study human biology and culture across vast gulfs of time and space through the dual lenses of science and humanism via the filter of archaeology to avoid excesses of soulless method and methodless soul per Watson (1991). Archaeology has done a lot of soul-searching, as has its parent anthropology, regarding basic theoretical premises and methods. A bit more reflection and thinking in regards to what we can and want to do with AGP is therefore perhaps necessary. What do *we* want it to do?

In answering such a query one must engage in a bit of sententiousness, perhaps even a bit of evasiveness. In one's opinion, we need to get away from our tendency to view AGP as just a way to find sites or to spot material underground or establish site limits. Those things are very useful but they don't even tell us very much about people unless they are embedded inside a larger anthropological theme of research. We should be asking instead how to incorporate AGP within anthropological research designs. How can we use AGP to answer questions we might have matters of anthropological relevance? We have a solid body of previous scholarship relating to scientific uses for AGP. We do not, in one's opinion, have nearly such a solid footing with regard to matters of actual humanistic concern, such as agency, ideology, style, or concern. Can we use AGP for evaluating such things? Which ones? To what degree? I would argue that we certainly can use AGP humanistically, through that collection of epistemology and technical methodologies known in archaeology as middle-range theory.

AGP IN ARCHAEOLOGY AS MIDDLE-RANGE RESEARCH

Middle-range theory (hereinafter MRT) was developed in sociology, and as with many concepts and techniques, archaeology appropriated it and remade it into something new (Raab and Goodyear 1984). In sociology, MRT refers to various approaches to the integration of research problems and data into bodies that can be arranged at different levels of generality. Those hierarchically arranged packages of problems and data can then be subsumed under larger general explanatory principles or "laws." A number of archaeologists, however (i.e., Binford, Schiffer), have substantially revised MRT in regards to its archaeological salience, such that the phrase has come to refer to a set of paradigmatic constructs and associated technical methods for inferring human behavior from material signatures, and for distinguishing between *anthropogenic* cultural processes and physical or chemical processes that condition archaeological material (Raab and Goodyear 1984). Binford, for example, (1977, 1981) uses MRT nearly interchangeably with "middle-range research" as a totalized set of mental and methodological approaches to linking data with inferences, (hereinafter MRR), whereas Schiffer (1972, 1976, 1987) described a "behavioral archaeology," essentially mechanisms for distinguishing between cultural and natural conditioners that affect site integrity. The main point to MRT is inferring human behavioral conditioning in archaeological material and in the segregation of cultural conditioning from other types. Finally, we should also observe that "theory" is a concept that has been seriously abused and contorted in archaeological epistemology, used to refer to nearly any hypothetical construct (Raab and Goodyear 1984).

It is pertinent to note that archaeology actually straddles the natural *and* social sciences simultaneously, and makes free use of numerous "theoretical" and methodological constructs from both categories. Archaeology is social science in its focus on humans and culture, while it is also a natural science in its extensive use of physical/material and environmental parameters (Kosso 1991). We combine humanistic concepts from the social sciences, such as sociology, psychology, economics, and history, along with material concepts from natural sciences including geology, anatomy, biology, genetics, chemistry, and physics. It is through the use of this conglomerate body of method and "theory" that archaeologists approach fundamental topics like site-formation processes, taphonomic processes, subsistence, and many others in attempts to distill evidence of human behavior from the archaeological record. In actual practice, a large constituent component of such research involves the construction of linking or bridging arguments that connect data with observations and inferences drawn from them. For example, in MRT/MMR site-formation processes are analyzed in terms of how archaeological sites are formed, by cultural, physical, chemical, and biological forces. Taphonomic studies include analyses of how formed sites are impacted syndepositionally and post-depositionally, by the same driving forces listed above.

Examples of MRT and MRR based on site-formation and site taphonomic agencies, in zooarchaeology and faunal analysis, and in lithic refitting and stone tool replication studies, are therefore quite traditional. In relation to site-formation and site taphonomy, there is also rather a large research field open to AGP: to assess newly located but unsampled sites (or, unsampled areas of previously excavated or surveyed sites) in terms of site integrity and taphonomy, prior to investing in costly excavations.

Moreover, in specific reference to CRM archaeology, AGP offers some extremely valuable opportunities for formulating new research designs and testing ideas about human behavior, provided that we establish a framework of epistemology, theory, and method that can be used effectively and comprehensively by CRM professionals. Contract archaeology has provided at least three benefits to the profession: by forcing researchers to cope theoretically and methodologically with heretofore unexplored and unexplained archaeological remains; by promoting a scientific merging of historical and prehistoric archaeology, and by stimulating archaeologists to probe the resource base in new and explicit ways for all possible dimensions of significance (Goodyear et al. 1978). There is no reason AGP applications cannot be routine in CRM projects. Given the tremendous potential AGP presents for subsurface surveying, in effect identifying areas worthy of time and capital investments, it is somewhat surprising that AGP is not already more a routine fixture in CRM.

AGP also provides CRM professionals with a tremendous advantage: AGP techniques are very good ways to prospect for unmarked burials. In fact, avoiding impacts to unmarked burials is one of the most important functions CRM performs, since their discovery at a site can rapidly bring activities to a standstill and require expensive stabilization efforts (Conyers 2006; Dionne et al. 2010). Routine, preliminary AGP surveys performed at the front ends of CRM projects would probably *reduce* overall monetary and labor costs by highlighting areas on which to focus attention. Areas indicating absences of archaeological material could be omitted from extensive testing and resources could be targeted to productive zones. Given the expense of many AGP surveys, many CRM firms and professionals would probably inform themselves about the techniques available to them, acquire the technology, and learn to apply them in the field if AGP were more an integral aspect of contract archaeology today. I would regard such a situation as inherently beneficial, since more archaeologists would presumably be using more AGP techniques, many by themselves instead of consulted experts, in more environmental contexts and at more site types. This would be unquestionably an informative transdisciplinary development, one that would benefit the profession as a whole.

While CRM professionals will need to determine the specific dynamics for greater inclusion of AGP in contract archaeology, several informative studies have provided us with a basic outline for how this could be accomplished. Kvamme and Ahler (2007) have described a comprehensive program of combining traditional excavations with resitivity, magnetometry, and ground penetrating radar for a multiyear project at the Double-Ditch Site, North Dakota. The Double-Ditch methodology could be a model of CRM activity at large, stratified sites. Jones and Munson (2005) presented a method for identifying the most likely productive areas of excavation emphasis at ephemeral sites, an obviously valuable addition to the literature and an understudied one that could also inform CRM projects. Lastly, in regards to the application of AGP at the Poverty Point site, Hargave et al. (2007) describe their procedures for using the technology to reveal features that are invisible to the naked eye.

CHAPTER 4

Toward Using AGP Anthropologically in Archaeology

PLACING AGP IN AN ARCHAEOLOGICAL FRAMEWORK

How do we actually situate or emplace AGP within anthropological archaeology? In an interesting recent paper addressing this topic, Thompson et al. (2011) suggest that our focus should be on situating or "embedding" AGP survey within an inquiry-driven framework. In such a framework, the questions that compel our activity actually drive the research process, so particular attention must be directed at formulating inquiries that AGP survey can be used to answer. Such questions can usefully be premised upon material aspects such as scale, configuration, orientation, distribution, and, to a lesser extent, frequency and associations of buried anthropogenic materials accumulated diachronically at what are termed "persistent places," portraying sensible instances of material change over time. In order further to provide a cognitive framework Thompson et al. (2011) suggest placing AGP within the context of established anthropological research projects. This is a sound concept, although in my opinion we should perhaps note that established lines of inquiry are not really our primary concern, but rather *establishing* them.

Archaeological sites are not really static phenomena, but are rather interactive collections of geologic, mineralogical, hydrologic, chemical, and atmospheric items associated in space and time, acted upon by physical forces and impacted by floral/faunal communities that exist within them; only some material at any site is actually conditioned by humans. Human behavior occurs at places, to be sure, but often as disjointed, finite episodes or vignettes across space and time, and as material indices of behavior enter the archaeological record we often lose a detailed sense of the behavioral interval over which they were deposited. Hence the present focus on small material scales. Thompson et al. (2011) also address the general absence of (presumably *anthropological*) theory in AGP application, attributing the absence to several factors: AGP (or remote sensing) is a "post hoc" technique applied as an add-on method and rarely published in AA journals; 2) sparse application of AGP to sites in the United States (vis-à-vis European sites); and 3) difficulty in using the equipment. We can't actually "see" human behavior via AGP; what we can detect are residual, material manifestations of behavior, or smaller-scale aspects of being, and more widespread knowledge of this among American archaeologists is crucial. We can, however, study some aspects of being through AGP by the employment of a middle-range research paradigm.

Due to didactic and pedagogical norms within American anthropological departments, for both undergraduate and graduate programs, it would probably be easier for American archaeologists to write about subjects such as agency and praxis than it is to write about AGP equipment settings and parameters and to explain the different results different such settings can elicit. Many of us were trained that way, with emphases on culture, culture processes, and human behavior. In the United States, it is normally common for archaeologists to interact just as much or more frequently with sociocultural anthropologists and linguists, or even sociologists, as it is for us to participate in interdisciplinary fashion with geologists, geomorphologists, soil scientists, and other specialists from different fields, including geophysicists and GAs: this is furthermore not a new development, as it is largely traditional (Gibbon 1984). Refreshingly, many American archaeology professionals are beginning or continuing to place their students in geology, geomorphology, and geophysics classes, and this new embrace of cross-discipline studies can be expected to elicit future results.

One does not, however, wish to suggest that we all peer at one another across some vast, impenetrable gulf, for that is not the case. We share many common reference frames, so formulating a new one for anthropological geophysics should not be beyond our grasp. Consider the following, for example. Increasingly, all archaeologists at all career levels are pervasively affected by the dominant social contexts in which we live and work (Trigger 1989). Something that has been quite concerning has been the degree to which anthropological archaeology, in pedagogy, policy and procedure, literary publication norms, and simply in general have for some time incorporated various notions and metrics from the U.S. corporate orthodoxy, regarding subjective concepts like "efficiency," "effectiveness," "complexity," and other things (Thompson 2012, 2014a). An entire subgenre of archaeologists has transposed such subjective capitalist terminology into their opinions about the relative cognitive complexity of long-dead humans (Henshilwood and Marean 2003; Stiner 1994, and too many others to cite) from among what are at best extremely discontinuous if not entirely absent data Speth 2004). Considering we live in an age in which corporate hegemony has been established over many other sociopolitical entities we should probably be highly aware of the ways in which we adopt concepts from what are essentially trickle-down business apologetics received from the pop-culture media. In that sense, one is quite surprised American archaeologists are not more attuned to AGP, especially when various techniques can be seen on several popular crime series on television. We should, perhaps, revisit also the "fear" motif, notable apparently within CRM, that AGP will put such professionals out of work. In our insecure economy, are these unreasonable fears? It may not be that AGP will ever negatively impact any branch of traditional archaeology, but the fears it could happen are perfectly rational. Given the likelihood that we all know someone, or multiple parties, who have left the field or who have been driven from it due to declining numbers of jobs such fears become very understandable.

In this way we can see that archaeology is actually highly susceptible to various social influences (Trigger 1989). Archaeological interpretations are often heavily colored by class biases and ivory tower elitism (especially attitudes toward junior and adjunct faculty). Such developments actually appear to become more pervasive over time, and ought to caution us to be acutely aware of how our experiences in the present condition our interpretations of the past (Trigger 1989). The general trend is for archaeology to reflect attitudes and biases common to establishment society: such as from business, economics, and finance, as well as other sciences. It is more than obvious that we need not only to be aware of our biases but to mitigate them. We also *affect* our own biases, while we also have the power to change them. So long as all interested parties are included in discussions, symposia, and conferences, we should be perfectly capable of forging a collaborative anthropological geophysics inside American archaeology.

In the United States, however, archaeology is also becoming increasingly biased *politically*, especially in the generally unstudied but stifling political nature of teaching, research, funding, and tenuring (McGuire 2008). Archaeology and even anthropology, the parent discipline, has always been multicontextual, inhabiting realms of theory, method, environment, technical and material analyses, chronometry, biology, statistics, and prospection (Gamble 2001). Strangely, however, we have adopted (or have been infected by) many ridiculous notions of "competition," "efficiency," and "budgetary relevance" straight out of the for-profit corporate world, largely in imitation of private-sector business norms. One is not referring merely to CRM firms, either, as most of us would admit that the behavior and activity of anthropology departments in American colleges and universities have taken a decidedly "corporate" turn in terms of operations and protocols, especially regarding budgetary "austerity" and elimination of tenured positions as normalcy. Such developments come with steep costs for our discipline, our departments, our colleagues, and mainly our students. Developing broad transdisciplinarity can therefore be seen as a form of job security.

WHAT *IS* ANTHROPOLOGICAL GEOPHYSICS?

We unfortunately lack a cognitive guide for envisioning an anthropological geophysics. What does the phrase even mean? We can remove the *geo* prefix from the *physics* stem. Is there an "anthropological physics"? What is it? How should we assess AGP in *anthropological* terms? In other words, what specifically do we need to do to provide anthropological standing or relevance to AGP applications? This is a significant question. One thinks we can dismiss several things immediately. First, "anthropological standing" is not a condition of using AGP as merely "add-on" appurtenances to archaeological projects. Nor can we establish such relevance simply by attaching spurts of anthropological theory into our putative collaborative AGP literature and discourse. Anthropological standing can be established by a humanistic focus on aspects of being and behaviors. We have to build appropriate reference frames collaboratively, through discourse and dialogue and mutuality. Such discourse and dialogue will contain just as much anthropology and cultural topics as it will wave dynamics and equipment parameters. Would geophysical archaeologists find anthropological significance in zooarchaeological publications premised upon minute explanations of density-mediated bone attrition or taphonomic equifinality? That isn't to demean the effort geophysical archaeologists bring to their craft, but more an admission of fact. One has never fielded a request from a GA to provide thick description of the phsyico-chemical properties of calcium hydroxyapatite (i.e., bone). This is to suggest that GAs will need to address *anthropological* theory in forging our new collaborative reference frames as well as geophysics. The AGP literature as it is now is seriously deficient in anthropology.

HOW DO ANTHROPOLOGISTS THINK ABOUT AGP DATA?

While interpretations of AGP subsurface data are often subjective, the data themselves actually represent objective information regarding material and physics in the subsurface (Mahar 2011). This is a very key point: for whereas AGP data are objective, the manners by which we analyze them, synthesize them for presentation, and interpret them are more subjective. For archaeologists working with scarce funds, why bother using AGP techniques that only offer subjective interpretations instead of simply doing cheaper core or auger tests? While we can know that the data we receive represent "real" phenomena beneath the ground surface we cannot, of course, be really certain just what our AGP data portray until they are sampled via some verification process whether via coring, auger testing, excavation, or all of the preceding, and then compared to other technical datasets. In reality, this is very similar to "actualistic" studies and ethnoarchaeology, and represents an area of theoretical overlap with humanistic anthropology, where assumed universal patterns of human behavior intrude into at least partial conditioning of sensible, measurable material phenomena. Archaeological manifestations of humanism would include techniques, methods, and theories for spotting elements of human behavior, or aspects of being, in the archaeological record—as able. This brings us to a question again: What human behavior is sensible via AGP techniques? If we don't ask the question we won't get any answers, and if we don't reduce interpretive subjectivity any answers we do get are likely to be rather ethereal.

However, the problems faced by archaeologists doing humanistic AGP and, say, lithic replication analyses are actually quite similar. What avenues toward the human condition, or at least aspects of behavior, are visible to AGP? We can answer this somewhat through appeal to lithic replication studies: what indices of behavior do replication studies elicit? Replication studies can serve as an analogue for anthropological AGP. We have first to assume that material indices of behavior will sometimes be visible to AGP. We must then assume some degree of behavioral similarity exists between the behaviors that served as proximate causes for material phenomena that appear the same or similar in AGP data. Our theoretical frame might be something like, "We know that humans sometimes do things at archaeological or ethnoarchaeological sites that result in the formation of a given feature or structure." We would then infer that humans doing similar types of things at different sites could result in things that appear the same or similar in AGP data. The preceding points to the need for what Thomas (1987) proposed as a bridge between AGP data and archaeological or anthropological theory. This "bridge" would consist of linking arguments and methodologies for connecting data with epistemology in an explanatory framework. Many of us know this component in our archaeological toolbox as middle-range theory.

Mahar (2011) suggests several strategies for this, including the multiple means approach (use of multiple, different AGP techniques in tandem at archaeological sites) and data integration (overlaying of different datasets produced from different AGP methods to produce composite imagery). Unfortunately, instances in which there is time, money, and personnel to conduct multiple-methods AGP anywhere may not be common. There exists a noteworthy body of such research (see the many such studies conducted under the purview of David Hurst Thomas at the American Museum of Natural History), in which combinations of AGP techniques have been applied at sites by collaborative, transdisciplinary research teams. These are, however, clearly exceptions and not rules, as seminal research ventures often are. Such research would come with daunting price tags and time commitments, just to gather the data. Years could then be spent processing, post-processing, and compiling data into usable imagery for further analyses. Yet, how many of us have the time, money, ability, and personnel to conduct such research? In an interval of severely shrinking research funding, acquiring money to buy, rent, or otherwise arrange the gear necessary for such undertakings would be considerable and infrequent. So the majority of AGP research will never approach that level of sophistication and thoroughness unless costs associated with it continue to decrease as increasing numbers of academic and CRM archaeologists learn how to use it.

In terms of implementing an AGP survey, it is important to trace the sequence of events that occur, noting each step and how they link to compose the entire process. For there is a process, in fact, there are *many* processes involved in applying AGP. There is no one-size-fits-all approach, and the actual sequence of events will vary based upon the AGP technique employed at a site. It may also be necessary to compose new methodologies based upon field conditions. In general, an AGP application would include the following steps: 1) Field reconnaissance and pedestrian survey, taking account of vegetation, lithology, sedimentology, and hydrology; 2) Emplacement of a survey grid or sampling array, depending upon the AGP technique employed; 3) preliminary AGP operation at the site to calibrate equipment, gauge appropriateness of sampling intervals, and arrays; 4) Performance of the AGP survey; 5) Back up raw data: Conyers (2012) suggests at least three separate copies; 6) Signal processing to transform raw data into graphics; 7) Further processing of data to produce images, maps, and profiles.

Practical experience should help to familiarize AAs with the dynamics of AGP applications, so approaches described in prior research and publications can be combined with and incorporated within new research frames developed by other archaeologists for their own purposes. This should prevent new practitioners from being forced to start from scratch, and adds the benefit of having a framework for developing their own unique research programs. The process places demands on the archaeologist, however, in order to maximize the potential of AGP in our research designs, in that we need to develop several types of general and specific knowledge and experience before even attempting to formulate an AGP research design. These can be broken into general and specific categories. In terms of general knowledge, one should be familiar with the AA and AGP professional literature, basic GA and AA methodologies (i.e., the basics of conducting an AGP survey, basics of faunal analysis, etc.), and archaeological reasoning. More specific examples would apply at the level of the site of proposed AGP application. Such specific knowledge and experience would include expertise at the regional level (i.e., the Southwest, the Midwest, the Paris Basin, etc.), at the level of the physical landscape, and

Table 4.1. General and Specific Knowledge and Experience

General Knowledge and Experience
Knowledge of AGP and AA literature
Knowledge of and Experience with AA and GA methodologies
Archaeology Theoretical knowledge*
Specific Knowledge and Experience
Knowledge and Experience at Regional level**
Knowledge and Experience at level of physical landscape
Knowledge and Experience with geomorphic setting, landforms

*History of Archaeological epistemology
**I.e., Southwest, Midwest, Paris Basin, etc.

with the geomorphology of landforms and the physical setting. This material is also presented in Table 4.1.

DEVELOPING A TOLERANCE FOR SUBJECTIVITY, AMBIGUITY, AND RELATIVITY AMONG GEOPHYSICISTS AND FOR GEOPHYSICAL AWARENESS AMONG ANTHROPOLOGICAL ARCHAEOLOGISTS

Use of AGP in archaeology involves considerable amounts of subjectivity, ambiguity, and relativity. It is not possible to ground verify what it is we think we see in AGP data, whether patterned or not, if excavations are not conducted afterward. There can obviously also be multiple explanations available to explicate whatever it is we think we "see" in such data. We must also, therefore, interpret what we see in images relativistically and comparativistically; our AGP data and images are not underground pictures, they are models. We have to use these models analogically in comparison/contrast with other phenomena, even though these models are not themselves analogues. It may, in fact, be rather rare that we are able to gain solid explanatory power in relation to buried archaeological materials solely through AGP techniques. Such data have to be compared and contrasted with other datasets and interpreted alongside other technical analyses of materials identified via AGP that are excavated later. With regard to archaeology, there are many ways of interpreting material, let alone EM reflections of them or anomalies located via EM and resistivity. How can we set about interpreting reflection and anomaly images of subsurface materials, based on the widespread use of modeling and analogy in archaeology?

While extremely common in archaeology, we sometimes treat analogies and models as if they were the same things, while they are not. Analogies, for example, involve estimates of similarity or comparability and contrast based upon observation and/or documentation. Models are more schematized descriptions of systems. We might, therefore, suggest that whereas models are actually more theoretical constructs, analogies are more heuristic devices

Table 4.2. Subsistence/Settlement Models.

Author	*Model*	*Derivation*	*Pub. Date*
Mortensen	Circulating/Radiating	theoretical/historical	1972
Wiessner	Population	theory/simulation	1974
Wood	Locational preference	ethnoarchaeology	1978
Hodder	Spatial/material	ethnography	1979
Binford	Foraging/Logistical Coll.	ethnoarchaeology	1980
Odell	Representative sampling	theoretical/statistical	1982
Chang	Fishing foragers	ethnoarchaeology	1988
Rocek	Seasonal economy	ethnoarchaeology	1988
Sullivan	Settlement chronology	ethnoarch./survey	1988
McCartney & Glass	Statistical/simulation	theory/simulation	1990
Headland	Socioeconomic	ethnography	1997
Kolb & Snead	Spatial/material	pedestrian survey	1997
Hewlett et al.	Cognitive	theoretical	2000
Prentiss & Chatters	Historical contigency	theoretical/historical	2003

based upon a process of comparison and contrast (Thompson 2014). Sumerian and Greek city-states can serve as analogous examples of ancient city-state political models. In similar fashion, Mycenae and Cuzco are often held to be analogous in terms of cyclopean wall construction methods. Arctic hunter-gatherers are often used as analogies of Upper Paleolithic Eurasian foragers. Archaeologists routinely use ethnoarchaeological analogies and actualistic research (for example, lithic replication studies) as analogues for past behaviors. But those are not really models, they are comparisons and contrasts. A set of models based on humans' subsistence is listed above in table 4.2 for reference.

Inspecting various of the models in Table 4.2 above, it is clear that many of them are analogous to one another, in the sense that they share similar common premises and structures. We can use models and analogies comparatively in AGP, and in fact we have been doing so for years: hence the presence of online databases of AGP images and results. What do such databases as yet lack, however? What do we need to add to our repertoire of AGP images?

Furthermore, use of even isolated AGP techniques can be complicated and both time- and labor-intensive, to say nothing of expensive. Significant amounts of expertise must be developed for interpreting AGP data as well, similar to learning techniques of musicianship, or writing, and other subjective creative skills. These are not linear pursuits, and there isn't really any "one best way" to do any of them. Nor is there likely any one best way to use AGP in archaeology. We have sets of techniques and methods that have been proven to work, at least in producing some imagery of intact subsurface archaeological materials, in some contexts, under certain conditions. Not everything works in all cases at all sites; this makes an awareness of the AGP scholarly literature crucial for the practitioner. Most of us have to make trade-offs every day, such as which AGP technique would get me the best results? How much data do I need to acquire? How tight do my sampling intervals (transects) need to be? How much ground am I covering with my survey? How much time do I have? How many people are helping? In many respects most of the other variables affecting the utility and practicality of an AGP survey are themselves dependent upon the biggest independent variable affecting outcomes: how much money do I have? Where can I find accessible catalogs or databases of "comparative collections" from previous AGP research?

With respect to actualistic studies in zooarchaeology and lithic replication, the presence of sizeable, variable, and representative comparative collections are simply mandatory. Without such references even the ablest faunal analysts might be reduced to considerable amounts of guesswork in zooarchaeology. Lacking comparative collections, the lithic analyst has little idea of the range of variability that stone implements can exhibit. Therefore, if we're going to use AGP in anthropological archaeology, we need a set of accessible AGP databases that is large enough, variable enough, and representative enough of intact subsurface archaeological material that we can make sound inferences from it. Such a body of images would help bridge the gap between AGP data and anthropological explanations of them, between material and possible behavioral conditioning of it. AGP images abound; what we really lack are widespread linking and bridging arguments constructed to explain the material patterning we either see or don't see in AGP data. Unfortunately, this is an aspect of the "successful" vs. "unsuccessful" dichotomy common in AGP publication norms. Unsuccessful images, or interpretively complicated ones, with multiple, equifinal proximate causes, tend not to get published.

AAs and other nongeophysicists must realize that GAs use a highly technical language of physics because AGP is not perfect, and because a wide variety of conditions and properties can effect outcomes, in ways that we need to rule out operator error, outside interferences, and multiple possible technical pitfalls. While the AGP literature can perhaps be difficult for humanistic archaeologists to understand, the reasons for the descriptive physics digressions are due to the power and sensitivity of the technology, and to the possibility of human error. The technicality is not a conspiracy to confound anthropologists. The technology always works (unless it's broken or unless serious outside interference occurs), while the contexts in which it is applied vary, primarily as a function of *who* is using the equipment, *where* they use it, and *how* they use it. To look at the matter from another angle, it is a serious stumbling block for the acceptability among GAs of AGP research conducted by AAs when semi-adepts fail to frame research "correctly," or at least comprehensively fail to describe what they are trying to do explicitly. More widespread availability of AGP courses at American universities and broader awareness of the utility of the techniques can improve the situation.

HOW CAN WE BRIDGE THE GAPS?

Where can we look for examples of how to bridge the gaps? Conyers (2012, 2013), Kvamme (2003, 2008), and Hurst Thomas et al. (2011) are particularly adept at negotiating the gaps between anthropological and geophysical archaeology. By bridging our disciplinary, literary, and communicative gaps there could be many innovations possible in AGP, and there could be augmented *social* and *cultural* interaction between disciplines. Within any academic discipline there exist various cultures, and they feature in-groups, out-groups, and many subcultures that condition the professional pursuits of the various practitioners, and it is this way that our fields echo the larger culture in which they are embedded and from which they draw their potency. One submits, therefore, that for anyone interested in participating in AGP research we have to expend considerable personal effort, disciplinary currency, time, labor, and money to inform ourselves in a greater "cross-cultural" depth to bridge the gaps between AA and GA, and this is especially true of more anthropological professionals.

SEEKING VIEWS OF PAST BEHAVIORS

In some sense, we are seeking AGP visibility of human behavioral traces, analogous to animal burrows, beaver dams, deer runs, and other alterations of materials situated within habitats that were driven by human cultural ecology. Human aspects of being involve material changes to locales, landscapes, and habitats, only some of which are geophysically visible. That brings us to the author's primary assumption, namely, that human behavior, or behaviorally conditioned aspects of the human habitat, can sometimes result in identifiable material signatures visible via AGP techniques. Humans often organize or modify their habitats in recognizable ways, often through engagement in ethnographically attested behaviors that result in allegedly diagnostic alterations of their habitat and patterning in excavated anthropogenic materials (Binford 1980). Another major assumption made explicitly is based upon the law of superposition, that for the purposes of this book in terms of AGP deeper material is usually older, barring evidence of material disturbance. In this way, changes in material distributions, scales, configurations, and densities that portray variability as a function of subsurface depth can be regarded as indices of behavioral change over time as well. Scalar differences between material items imaged via AGP could be associated with other differential properties, such as wealth, prestige, power, and population. Changes in material distribution or density and frequency could indicate changes in labor organization or resource acquisition, which are also based on other variables. AGP data need to be linked with such interpretations. Until subsurface features detected via AGP are actually ground verified, we can't actually say what they are. Even massive buried things with relatively clear "signatures" are only basically just apparitions. Subjectivity comes with the territory.

PATTERN RECOGNITION

Much archaeology has traditionally been heavily involved in various forms of pattern recognition (PR) programs (Binford 1981; Hietala and Stevens 1977; South 1978). The assorted "patterns" archaeologists seek and study are widely varied, in some instances reflecting particularities of both theory and method, but also evincing connections to less obvious aspects of data, while there are also material patterns, spatial patterns, temporal patterns, statistical patterning, methodological patterning, systemic patterning, chemical patterns, and settlement patterns (Hardy-Smith and Edwards 2004). The common (ironically patterned themselves) elements in such research are a focus on detecting, analyzing, documenting, and, in some cases, predicting an assortment of patterns. These patterns can be more or less obvious, observed directly in patterned landscapes, or material distributions of artifacts, for example, a lithic or faunal assemblage at a site exposed by excavations. Yet, some patterns may be visible only in very high-order data analysis, indirectly observable except by the employment of sophisticated technology and data processing methods, such as in cluster or nearest-neighbor analyses.

It may, however, be useful briefly to step outside the archaeological literature to examine, at least cursorily, some other uses of pattern recognition procedures. There is, for example, a scholarly journal, appropriately entitled *Pattern Recognition*, devoted entirely to detailing research methodologies for recognizing various patterns. In the journal *Pattern Recognition*, a dizzying number and variety of scholars from disciplines ranging from anthropology, to computer science, to air traffic control, to military strategy offer for review their contributions to pattern analysis. One such article, by Martinez and Hamsici (2008), uses a computerized discriminant analysis of cranial morphology to assess hypothetical evolutionary models in an attempt taxonomically to classify the Flores hominid LB1 (as an offshoot of *Homo erectus*). In an earlier volume of *Pattern Recognition*, Ozawa (1990) uses various sophisticated mathematical and statistical methods to analyze patterning in the development and perception of optical illusions. Finally, in an even earlier volume of this journal, Tang and Huang (1980) demonstrate a usage of a linguistically derived syntax-semantics method to identify various kinds of aircraft, from photographs and from different angles and perspectives. Other common uses of PR include facial recognition programs, fingerprint matching, data mining, and similar large-scale data analytical applications.

What should we call patterns that we think we recognize in AGP data? Have we developed a handy lexicon of terminology specific to the things we see or think we see in such images? The geophysics literature provides some terms, such as anomalies, reflections, and refractions based upon wave dynamics and basic observation. Assuming we detect a material pattern in our archaeological AGP data that we can recognize as an actual pattern, what nomenclature do we use for it? Since there is not a reference set of AGP patterns waiting for us, we have to use our own language, our own descriptions of what we see. Descriptions of subsurface patterning will thereby be subjective. Since the patterns we see in AGP data will be in various shapes and sizes, we should discuss the scale, orientation, and configuration of such patterns in terms relevant to the site. What is the site scale? Is the pattern specific to particular area of a site (intrasite variability) or the whole? Is it something that has only been identified at that site (site-specific)? Does it characterize the site or is it spatially anomalous? Is it diagnostic at the level of the landscape, vis-à-vis other sites? How should we compare patterns? Particularistically or in generalized abstract fashion?

Is pattern recognition anthropological? That is a peculiar question, and one would answer it by noting that sociocultural anthropologists, medical anthropologists, and other anthropologists fortunate enough to research living people do indeed study patterns and employ pattern recognition techniques. Human behavior is largely patterned, especially in terms of economics, kinship, religion, subsistence, and numerous other cultural phenomena (Nanda and Warms 2006). Such patterned behaviors certainly lend themselves to analysis and inquiry. In similar fashion, multiple kinds of subsurface material patterning interest us as archaeologists. Patterns figure very largely indeed in all subdisciplines of anthropology. Themes to which we should accustom ourselves: Change over time, change in material scale, configuration, distribution, and density over variance in depth, change and differential use across space, and looking for other aspects of patterned variations. It's impossible to delimit exactly what we might expect to find, especially given what we don't know about undiscovered archaeological sites.

PATTERN RECOGNITION AND PATTERN IDENTIFICATION

Humans are by virtue of being human adept at pattern recognition, not only within the bounds of our five primary senses. Many other animals are also adept at recognizing patterns: primarily visual (but not limited therein) via coloring and oppositely via deceptive camouflage, described as signal architecture (Rosenthal 2007). Why? Because patterns exist in nature. They abound.

We see things and, having analog minds, we unconsciously put them into drop-down boxes that correspond to what we already know. Patterning exists in the distribution of floral and faunal communities as functions of regional variation in latitude, altitude, hydrology, soil, and other resources (Willig et al. 2003). But just because we can recognize the existence of a pattern, does that mean we can identify it or understand what caused it?

In machine language and through artificial intelligence it has been demonstrated that mathematical pattern variations result in perceptual discriminations by observers based on convergence and divergence (Wurtele 1965). Like patterns result in convergent perception while unlike patterns result in divergent perception (Wurtele 1965). In other words, patterns vary by means of being like or unlike other patterns or things, in analog fashion, even if composed of digital information, which is how our minds function, in analog. Since this associational quality accrues to perceived patterns according to how our minds work, a particularly noteworthy aspect of our pattern recognition system involves learning, or a learning curve, such that agents display a decrease in identification errors as a function of duration in pattern recognition training (Albert 1963). We learn by absorbing new things and categorizing them in relation to things we already know.

Binford's (2001) paper on research problems suggests that pattern recognition studies of archaeological datasets should offer fruitful opportunities to generate theory. What sort of patterns do archaeologists generally recognize? Where do we find patterning? We unquestionably find patterning in material distributed at sites, as well as in the distributions and locations of sites themselves. We also observe patterning in space, analyzing the spatial relationships between things (including absences of things). We attempt to see social patterning (physical distributions as well as more "cultural" locations) in many things, as well as behavioral patterning. "Well, what else?" we may reasonably ask. Archaeologists frequently recognize patterns in data, revealed through a wide range of analytical and statistical methods. We recognize certain historical patterns relating to the origin and sophistication of archaeological methods and theories, for example, as well as recurrent successions of paradigmatic preferences and all manners of epistemological excrescences. In fact, one could probably find as many possible definitions for archaeological pattern recognition as there are archaeologists.

South (1978:223) described in simple and elegant terms a very complex and multifaceted process: "In order to identify behavior and process reflected in the archaeological record the archaeologist must concern himself with pattern recognition using all data sets at his disposal." Not only that, but we're also advised by Hietala and Stevens (1977:539) that, "The suggested procedures (. . . of pattern recognition . . .) fall into three conceptual areas of methodological investigation, corresponding to the problems of pattern detection, pattern estimation, and pattern prediction." A reasonable APR process would then seem to include multiple sources of, and expose them to manifold analytical procedures in order to locate, evaluate, and then predict patterns based on the evaluations.

Pattern recognition in AGP is obviously related to the theoretical questions and research frames we establish. How does archaeological theory impact PR? Odell (1996) suggests that archaeological theory is a set of systematically connected and verifiable propositions concerning the underlying principles that condition relations within a set of phenomena. Since there is no "archaeological grand theory" in the vein of gravity, relativity, or natural selection, we approach theory in archaeology by formulating and testing models. Models are *patterned.* We can consider AGP imagery of subsurface phenomena as our models; models we create through AGP survey (data collection), the various settings and parameters we use to calibrate our equipment and signal-processing software (data analysis), as well as final AGP images of the subsurface (data presentation). Some portion of these AGP models are based upon actual material distributions, but the manner in which we see them depends very greatly upon how we adjust equipment and computer software. So these AGP models aren't really purely cognitive models. In our case, we should ask what material patterning is indicated by our AGP data and how does it bear on our research goals. We have to recall that AGP is remote sensing, not underground photography, so there will be some unavoidable interpretive ambiguity in any patterns we might discern.

Yet, identifying patterns, finding noteworthy relationships (or absences thereof) between things and quantities, are really not the end in itself. What does it mean if we find no patterning at all? What does it mean when we find many? Does the pattern(s) have some intrinsic meaning of their own, some independent identity beyond our interests and models? Not always, certainly, and not until excavations define the characteristics of modeled material patterns. Thus, the process we use is a scheme of recognizing patterning and then *relating* our pattern(s) to something(s) else, such as human behaviors, which strangely enough often consist of or otherwise feature yet other, deeper patterns. We identify patterns through comparison and contrast with other patterns, or patterned absences, and we do this through our use of that most appropriate, archaeological account of argumentative affinity: *analogy*. We use patterns analogically, relating them and comparing/contrasting them with each other, as well as testing them against more cerebral constructs, such as biases, epistemologies, and even hermeneutics or ontology. The point is that archaeologists use patterns analogously as opposed to using them in isolation. Archaeologically speaking, a pattern may have no meaning whatever unless and until it is compared analogically with some other pattern, and this holds true regardless of which type of patterning we are concerned with (i.e., data, material, social, etc.). The point is primarily to teach ourselves how to identify anthropogenic patterns geophysically.

What patterns can we expect to recognize in AGP data and imagery? This is an enormous question, perhaps enough for an entire series of books. We cannot answer this, at least not with some finite range of possibilities or expectations. Not all sites have been surveyed geophysically, so we remain entirely ignorant about patterning we might find at them. For such sites (North American Archaic sites, for example) experimental simulation could be used to form a comparative database, allowing us to learn something about the sort of patterning we could expect in subsurface materials. So then we might ask, what *types* of patterns can we expect to recognize in AGP data and imagery, and how and according to what criteria would these vary? That is rather a different question, one that we can at least start to answer. Several kinds of patterning discernable in AGP data are: 1) Nonanthropogenic patterning of material at nonarchaeological sites; 2) Nonanthropogenic patterning of material at archaeological sites; 3) Anthropogenic patterning of material at archaeological sites no longer related to past human behaviors due to postdepositional disturbances; and 4) Anthropogenic patterning of material at archaeological sites indicative of past human behaviors.

We require the largest and most comprehensive "comparative collection" of AGP images we can assemble or acquire, capturing actual subsurface archaeological materials, simulated artificial archaeological materials, and blatantly nonarchaeological features. This ought to be also an open-source program, with multiple databases to which professionals could voluntarily upload their own content in order to increase this collection's range of variation. Larry Conyers, for instance, has some fantastic web resources, freely available and accessible for all. In one's opinion we should avoid "image hoarding" and miserly attitudes toward data. The archaeological record belongs either to nobody at all or to all of us. Much of this content, however, is "owned" by various scholarly journals, and they probably won't like freely shared copies of published images, and in any event published imagery is available. Especially unpublished data and images could be made more widely available. This would also especially be a very valuable opportunity to learn from "unsuccessful" AGP applications, and analyses of such data could be undertaken collaboratively by numerous parties.

INTERPRETIVE COMPLICATIONS

As Storey and Whittaker (2005) note, it is ultimately only through ground verification (excavation or other direct testing) that the nature of geophysical reflections off subsurface material can be verified or "truthed." In other words, until material is excavated, it basically exists in somewhat of a limbo state, as purely image-based media, neither "true" nor "false," almost like Schrödinger's cat (Wheeler and Zurek 1983). That said, however, there are some key advantages that AGP can offer in spite of the need for ground-truthing. For archaeological purposes, we might look at GP as a composite toolkit, a set of technological devices, user methods, storage media, late-stage data processing techniques, and final imagery for answering subsurface questions, each stage being important. This is similar to the ways in which pedestrian survey, auger tests, and GIS are used in archaeology. AGP could be, however, so much more. I would submit that AGP offers unique advantages in that once collected, AGP data can be manipulated infinite times since it exists purely in the digital domain (we will return to this in much greater depth in subsequent chapters). Sites can only be excavated once in traditional terms. That is it: once a site is excavated and retired materially, its story as an episode of human phenomena is basically over. The acquired data from a site can be reanalyzed, but it will never exist again as a composite assortment of material in spatial associations again.

Despite interpretive complications of the raw and even processed data, AGP can be both an entire survey method and a high-resolution "probe" of distinct smaller areas within a site setting; it has combined landscape, intersite, and intrasite applicability. It can be a survey method like pedestrian survey for prospection or a kind of electromagnetic probe, like an auger, for extremely fine-scale sampling. This functionality offers considerable advantages over pedestrian survey and auger-testing because of the rapidity with which AGP data can be collected (Leckebusch 2003). In many cases, an AGP survey can be done *faster* than a pedestrian one, noninvasively at that in contrast with auger- or shovel-testing, and the data archived and manipulated indefinitely. The primary issue is, however, subjectivity of AGP data interpretations and their salience to excavated materials; without ground verification, AGP models are basically untestable. This is a strong argument for a comprehensive comparative collection of images across a wide range of variation.

What limits are there to anthropological uses of AGP? Where are the boundaries and how do we find them? For example, AGP may never be able to answer many specific questions premised upon understandings of kinship networks. Yet, AGP probably can answer a lot of questions premised upon understandings of changes in material scale, orientation, and distribution over time during and between, for example, and between Paleoindian, Archaic, and subsequent periods. However, unless anthropological archaeologists demand more from AGP, and participate more in implementing it, we won't get more. Prospection for sites is actually only one component of archaeology, and it is not by itself anthropology. A focus merely on trying to image subsurface archaeological material might be informative, but it isn't explanatory unless linked to actual behavior through middle-range research. The situation is somewhat akin to early to middle twentieth-century archaeology in an intense preoccupation with purely material analyses; where documenting and categorizing various material phenomena were held to be the goals instead of trying to understand changes in human culture that *caused* changes in artifact styles. In many ways, the AGP literature now reflects trait lists compiled from "successful" testing of sites with a cataloguing of physical parameters that were conducive to the successes.

EXPERIMENTATION WILL BE CRUCIAL

A very important and ongoing battery of organized experimental ground-penetrating radar (GPR) studies led by Larry Conyers of the University of Denver (in collaboration with the U.S. Defense Department's Strategic Environmental Research and Development Program, SERDP) began to catalog variability among archaeological GPR applications (Conyers 2004). This long-term research project has advanced archaeological GPR research drastically. Yet, there still remains a virtually limitless universe of potential experimental studies in AGP. Particularly useful would be continued use of multiple integrated AGP methods used in different types of sediment matrices under controlled conditions of temperature, moisture, and even other nonvolatile or nonpolluting liquids to ascertain the effects of such physical conditions and materials on AGP imaging. Another promising area of future AGP research involves geophysical archaeology at older, more ephemeral sites, from the Paleolithic through the Mesolithic/Epipaleolithic. Such research may also not involve merely using AGP at more sites, but using it in different ways, in order to push the technology and methodology to "see" more than is normally expected.

Yet, we must be cautious. One seriously doubts that AGP will ever reach the point where images generated can be used simply as underground "photographs." Some of the conversations one has had with anthropological archaeologists suggest that underground pictures are exactly what they want or expect. Even when an experimental database of images is compiled, showing how numerous archaeological phenomena appear in images, interpretive complications will remain. For example, it would be a very bad idea to hold up actual photographs of Paleolithic or other small-scale, ephemeral material from one site, then conduct AGP surveys in other areas, and then simply to expect the AGP images to look just like the photos. Yet, it sometimes seems as if archaeologists unfamiliar with AGP might expect this. Experiments have shown that AGP images of subsurface material can look quite different than excavated archaeological materials or photographs of the same things

(Leckebusch and Peikert 2001; Thompson and Storey n.d.; Thompson 2011). Why is this? Differences should be expected between exposed archaeological materials and GPR images: physical and energetic properties of electromagnetism explain why we should not, except through ignorance, expect identity between excavations, photographs, and radar images. This is also a significant argument in favor of experimental AGP research on simulated archaeological materials (Thompson 2011; 2014; Thompson and Storey n.d.).

AGP data are collected and recorded nearly instantaneously while excavations progress over intervals of time. AGP waves propagate at different speeds and wavelengths relative to the method employed, whereas excavations and photographs require visible light. Excavations completely disaggregate sediment, organic material, associated artifacts, and moisture slowly through time; radar surveys collect and record data reflected from sediment, organic material, associated artifacts, and moisture all at once. Furthermore, clods and clumps of soil vary across a wide range of properties according to how they are formed, whether they clump on bones, lithics, rocks, or other materials. The moisture content, for example, will slightly vary between identical clods or clumps of soil based upon whether they have clumped on bone, flint/chert, or FCR. Variations, even slight, in moisture content profoundly affect radar reflections off materials. Other variables affecting this dichotomy are the numbers and sizes of interstices between sediment grain; bulk density, and packing (compaction) of sediment.

Should we expect AGP imagery of subsurface material really to resemble photographs at all? This is a cogent question, and one that requires some attention. First, we should remember that the subsurface is not "empty"; it is, instead, "full," full of sediment, moisture, and archaeological deposits. Archaeologists might label a site as "empty" if it contains no archaeological material, but it is not "empty" in fact. This highlights an aspect of our own cognition and discourse. We are rather often not as careful with terminology and vocabulary as we ought to be. A truly "empty" site would contain nothing at all, not even gases, and would be in effect a vacuum. Yet, AGP cannot actually "identify" what we might describe as an "empty" area from a "nonempty" one containing artifacts: all AGP can show us are the recordings of reflected signals transmitted from the surface downward. *We* have to identify the things AGP might show us, or at least try. AGP is analogous to photography, in the sense that both elicit images of reflected waves from GPR and anomalies in magnetic and electric methods. Just what those waves reflect off, and what anomalies portray, varies by method, geology, hydrology, amounts of electromagnetic interference, solar radiation, terrestrial radiation, and many other local or even point-specific variables. Again we feel that it is perhaps most productive to view AGP images, even the best ones, as merely models of underground stuff, not pictures of it.

Along with the causes of variability noted above, AGP imagery is also conditioned by sampling intervals, or by grid transect spacing during data collection phases. If using an EM method, equipment arrays also have effects upon final image resolution. With ground-penetrating radar, for example, higher frequency and resolution is offset by decreases in depth of wave penetration in the subsurface, so there may be inherent trade-offs involved. Yet, there are other reasons why AGP images can differ often dramatically from photographs, driven not by EM wave mechanics or sampling scales, but by mathematical properties of data smoothing and data interpolation by computer software. AGP data are not snapshots of buried things. They are, rather, computer files, packages of mainly textual information noting spatial coordinates and differences in subsurface wave penetrability and reflectivity. Computer software, such as GIS, "interprets" such textual data and assembles maps. Showing differences between reflections or anomalies requires human adjustments of signal-processing software, a detailed study of which is simply precluded for reasons of space. There are technical manuals for guiding such applications, as well as available online resources for more technical information.[1]

Yet, we might ask just what pattern recognition, pattern identification, experimental research design, sediments, and moisture have really to do with anthropology or even with humans in general. The answer, of course, is that they don't. So we might then wish to ask just how useful further experimentation with AGP can prove to be. Archaeological research designs are fabricated not to study people but to elicit understandings of the archaeological record, which exists in and consists of sediments and soils, varying degrees of moisture, and anthropogenic material, all in variable spatial association. In other words, as with archaeology in general, AGP's actual subject matter concerns static material things and various patterns of matter as opposed to dynamic human behavior (Binford 1983). It is the purview of archaeologists to provide epistemic linkages, between observed human behaviors and the material patterns they are observed to cause, and also between observed material patterning and the archaeological record. AGP can obviously assist in locating material patterning at archaeological sites and in site prospection. Archaeological sites are not randomly distributed across the landscape. Sites were located due to past contingencies. So what sorts of contingencies might be visible to AGP? Examples are provided below.

NOTE

1. For example, Larry Conyers's personal webpage: http://mysite.du.edu/~lconyer/; David Hurst Thomas's collaborative webpage at http://digitallibrary.amnh.org/dspace/handle/2246/6435; and the University of Arkansas's National Database of Archaeological Geophysics Results: http://www.cast.uark.edu/nadag/otherlinks/links_submenu.htm.

CHAPTER 5

Introduction to Research Frames

Relations between Research Procedures and Research Goals

In Binford's (2001:669) article on research frames is stressed the necessity of rationality of (research) procedure relative to one's goals. Whereas this may seem to be a simple tenet of any well-rounded research program, the combination of anthropological, archaeological, and geophysical versions of ontology render matters more opaque. For example, there is some tension between research goals and approaches from scientific versus humanistic archaeology; scientific approaches seek to construct a cumulative learning trajectory, or at least to reduce our ignorance of various phenomena while humanistic approaches seek to reconstruct past lifeways and to render virtually a faint shade of an emic observer perspective through the actualistic construction of ethnoarchaeological analogies (i.e., "method") and more theoretically based systemic models (i.e., "theory") (Binford 2001:669, 672; Thompson 2011:7). AGP, however, is neither purely geophysics nor archaeology, nor is it anthropology, so one could cogently ask just how to provide AGP anthropological framing. In other words, just how can we reconcile anthropology's humanistic aspects with geophysics' more purely scientific posture via a mediating archaeological paradigm?

Prior to detailing specific examples of material signatures of past human behavior and what we can possibly learn from them below, we should review some key points. The contemporary AGP literature bias of publishing successes at (mainly) massive sites results in general failure to account for failures and to learn from them and a large-scale bias in the literature. When "failures" are left unconsidered, we are robbed of deeper understandings of both relative successes and failures. We considered how AGP needs a systematic, comprehensive comparative database of imaging successes *and* failures, as well as published accounts of the differential physical conditions that resulted in both success or failure. This could also include some kind of before/after or side-by-side comparison of what is actually in the ground and what is actually imaged or not imaged.

Instead, we now have a voluminous literature of "we tried this in the field, it worked, and here are some images." This is not to impugn at all the quality of the research that has been done, for it is all very important. We have now to integrate AGP data with anthropological epistemology, a much more difficult proposition. Few real linkages have yet been made between AGP data and anthropological theory. A significant portion of the AGP primary literature is also devoted to exhaustive descriptions of equipment settings, operating parameters, and other thick description. To use a common archaeological refrain, most AGP research is stuck in "soulless method," relentlessly stressing methodology above other concerns (although Larry Conyers's, Kvamme's, Hurst Thomas's, and V. D. Thompson's thorough and ongoing and *open-source* collaborative AGP work are welcome additions). Why integrating AGP inside anthropological archaeology is complicated, however, relates essentially to qualities having to do with uncertainty: uncertainty in attributing definitive identifications to reflections from subsurface material; uncertainty in really explaining the causes of subsurface material effects prior to ground verification of what materials actually are; uncertainty in what AGP images actually *show* us (being visual models of subsurface material, not underground photographs).

Human phenomena do not occur inside a vacuum. There is a temporal, timing aspect to everything we do, and it is a very common refrain to observe that timing is everything. Anthropological archaeologists have therefore devoted countless hours to development of both relative and absolute dating schemes in order to establish specific chronologies at multiple levels of the site, region, and continent, and in some cases the entire planet (Hayden 1993; Renfrew and Bahn 1991). Despite many technological advances in absolute radiometric dating of archaeological sites, however, we are much less informed about the geographic or spatial loci of human behavior in archaeology (Binford 1983). This is true for a very simple reason: archaeological sites never preserve everything intact materially (Flannery 1982) and even more problematic is that the actual spatial, structural, and functional relationships between sites formerly associated through human agency are usually completely lost to us (Binford 1983). Yet, we know that archaeological sites were once located in human, or social, as well as geographic landscapes. The passage of time, the operation of monotonously constant physical processes of erosion, deposition, decay, and repositioning, and all manners of human activity (construction, engineering, etc.) obliterate ties between artifacts and between sites.

We considered above a very fundamental reason to undertake AGP in archaeology: the ability of noninvasively sampling space rapidly in order to try to establish a reconstruction of former spatial aspects of human phenomena that may be preserved in the subsurface material. We cannot excavate everywhere and expose the landscapes that once provided the physical setting for past human behaviors. We can, however, sample wide expanses of geography through AGP, without destroying precious in situ materials and rapidly across expanses of space. All the more, through AGP we can do this *accurately* and much less subjectively than through pedestrian survey. The best walkabout survey, with the tightest, straightest survey lines and transects, will never tell us anything about what is in the ground. AGP can give us a view into the ground, providing views to the past, economically and quickly. No matter how sophisticated becomes the AGP technology and application methodology, no geophysical survey gives a view *of* the past, because what it shows us is not actually material but is instead models of reflections or anomalies produced from material and material interfaces, differences between things, processed through computer software.

What do we do in our lives that we can expect to leave traces in the archaeological record, in the material inventory that future diggers will uncover? How many things do we do in our lives that leave traces in material and time and space that a photograph can really capture? With that in mind, what sorts of things do we do that future AGP could "see"? This is not merely a rhetorical exercise. We arrive at the main premise of this book: what did past people do in the course of living their lives that left geophysically detectable traces? How should we answer such an immense question? Above we considered various cognitive mechanisms to guide our research, including some (but by no means all) methods by which to guide our thinking before we ever set foot in the field. We considered theoretical and methodological processes for framing AGP research as an inquiry driven, middle range process, in which our a priori assumptions and the questions we pose, to ourselves, to our sites, to the different AGP techniques, to our data-processing methods, to the presentation of data in AGP imagery, and the interpretation of that data, all condition our conclusions. One argued that the final relative quality of the research we produce will be directly related to the quality of the questions we ask at the outset of our projects. One also argued above that we can improve the comprehensiveness of our comparative databases by engaging in more experimental instances of simulated, small-scale ephemeral occupations to accompany the extant collections of results from large-scale permanent sites.

Above we also discussed how anthropology, archaeology, and geophysics admittedly arrive at a search for patterning according to different methodologies. The respective common methodologies in each category have been developed for very different needs, treating very different subject matters and often radically variant analytical scales, according to very different variables. One argued above that common ground for anthropology, archaeology, and geophysics exists and that we need to seek it in order to address anthropological inquiries answerable via geophysics. It was suggested that transdisciplinary collaboration is necessary to find common ground and that theoretical, methodological, and participatory input is simultaneously required by anthropological, archaeological, and geophysical parties to achieve it.

Below one will present some examples and in-depth descriptions of ground-verified iterations of the four patterns above. In general, we would expect Type 1 patterns to include actual lithostratigraphic bedrock and unconsolidated sediment units, soil strata, frost heaves, and other phenomena as well as various physical disturbances and perturbations of such things by geomorphic processes. Patterns of Type 2 would include aspects of Type 1 that are found within the boundaries of archaeological sites, such as fluvial incisions and debris flows not directly related to human occupations but occurring near them or possibly impacting the archaeological deposits. Type 3 patterns would indicate the presence of actual material deposits primarily conditioned by past human behavior that is visible to AGP techniques. Patterns of Type 4 would represent the activities and impacts of contemporary or recent humans upon the material configurations of archaeological sites, such as excavated and backfilled test units, exposures, and trenching intrusive to a site. Obviously, most sites will not contain examples of each of these four types of material patterning, nor should we expect that. There are, however, some sites that do feature all of the above four types of material patterning, and this volume will henceforth consist of elucidating one site that does (the Magdalenian site of Verberie le Buisson-Campin, in Picardy, France, hereinafter VBC) and one site that doesn't (an Archaic occupation in Dubuque County, Iowa, United States), in comparison/contrast format.

The two sites one will compare and contrast are each located adjacent to contemporary incised fluvial channels. These streams are both situated inside carbonate bedrock drainage basins. Both of the two sites have been tested on multiple occasions via pedestrian survey, coring, emplacement of test units, trenching, excavation of lateral exposures and subsurfaces, and geophysical survey (Enloe et al. 2007; Rogers 2008). Both sites have produced diagnostic lithic assemblages. The results from the two studies, however, differ in that one (Verberie) was relatively more "successful" (resulting in the gathering of data conducive to the productive of imagery useful for directing subsequent excavations and anthropological analyses) whereas the other (13DB674 in Dubuque, Iowa) was more "unsuccessful" in the above regards.

What do we want AGP to do? As will be described below, AGP can be used to image submeter-sized archaeological material configurations in enough detail that, using a comparative image collection, one can infer some human behavioral conditioning on the material. What would this conditioning be? One will refer to the different material signatures of different human behaviors, and perhaps differential seasonality in site-occupation scheduling; perhaps the relative size of human occupations; perhaps even some index of the frequency of human occupations. How can we get to where one wishes to go? Below we will see how a comparative collection of small-scale AGP images indicative of what is found at some open-air sites can be applied an inquiry-driven research program. Below we will see an approach for integrating AGP data along with faunal data and formulating a multidisciplinary methodology for linking AGP data with past human behavior. The keys to future such integrative studies will be the depth and breadth of comparative AGP data and image collections across a wide range of variation in simulated and actual subsurface material patterning. We discussed above the need to consider such phenomena across a range of variation because we cannot know the complete morphological diversity of ephemeral site features, nor do we know the complete range of configurations, scales, orientations, distributions, or frequencies. Given the incompleteness of our data, broadening our awareness of the potential range of variation ephemeral sites could exhibit is perhaps best addressed through the integration of experimental simulations to a greater degree than we have at present.

CHAPTER 6

Success at Verberie

WHAT IS "SPACE"? WHAT IS "PLACE"? HUMANS ARE NATURAL ASPECTS OF planetary environments. There is no absolute boundary between the human world and the "natural" world. Humans modify their own environments and their own ideas about their places in them. Can we freely exist "outside" the environment? Nothing is static. Stasis is an illusion caused by inadequate sampling. If we wait long enough everything changes over time. Everything evolves. Humanity and its ancestors have existed for at least 7 million years. For about 4.5 million years of that interval humans and our ancestors evolved biologically to environmental change. For the last 2.5 million we have increasingly adapted *culturally* to external change. Humans now simultaneously occupy a cultural, as well as an environmental, niche. What advantages/disadvantages can you imagine for either niche?

What is space? Space is context or setting. Space is area. Space is dimensional (x, y, z). Space is where everything happens. All events have a "place" where they occur. We can imagine space as the place where things happen. Space is not inherently cultural; it exists in its own right. What is a place? Place can be a particular point or location in space. Place can include the human and physical characteristics of that location. Place includes human modifications, investments, traditions, activities. Space contains place; place contains us: space can be the context of place; place can be our context. Space is not a human construct whereas place is. A physical place includes the physical characteristics of a location. Rivers, forests, rock outcrops, hills, valleys, mountains, and other phenomena can characterize a physical place. What things characterize cultural places? What criteria should we use?

In this chapter we will approach some of the actual methodology and applications involved with the use of AGP in anthropological archaeology. It will describe the way my colleagues and I developed several methods for experimentally calibrating a ground-penetrating radar survey through the use of artificial "sites" (i.e., lithic knapping products, animal bones from various species, and carbonate rocks) in a sandbox, in an actual in-ground setting, and field applications of GPR to two different low-density open-air sites. Although some folks have applied AGP to prehistoric open-air sites, the applications have been few and far between, and the results have been indecisive at best. Due to this, my colleagues and I set about designing an experimental procedure to test AGP technology (in this case ground-penetrating radar) as well as *ourselves* and our assumptions, methods, and conclusions pertaining to application at low-density open-air sites.

Thompson and Storey (2010:3) observe that: Whether prospecting for "new" sites or imaging known sites, the basic interpretive program involved—when stripped of technological and methodological overlays—is pattern recognition. But with few published "successful" applications of AGP to low-density, open-air Paleolithic sites, what patterns could we expect? How would reflections off the kinds of subsurface archaeological material from open-air Paleolithic sites look in relation to the same materials when excavated and exposed in photographs? Another way of saying this, as well as highlighting a particularly vexing interpretive issue, is to ask: in what way can we distinguish between reflections or anomalies generated by legitimate archaeological materials, and those generated by nonarchaeological materials?

The question, here, then is what sorts of patterns in material scale and configuration are even visible or observable via AGP? Whereas some experimental and practical applications suggest very high-resolution capabilities for certain AGP techniques, I describe an attempt to build on such studies in such a way as to identify complex subsurface material patterns at the French Magdalenian site of Verberie le Buisson Campin (VBC) and a North American Archaic site in Iowa. What sorts of subsurface patterns are visible to AGP and how do they look? Primarily, such patterns seem to consist of various geometric arrangements of material, some sedimentological, some hydrological, some archaeological (Thompson 2011; Thompson and Storey 2010). Familiar shapes seem to be the key, such as lines, circles, and polyhedrons, as well as patterned absences (voids) in material and spatial configurations, in plan view and profile. Some shapes will conform to natural soil horizons and lithostratigraphy. Some such shapes will conform to nothing visualized before and will be totally unique. Identifying patterns that conform to likely archaeological materials instead of various nonanthropogenic processes and structures is a skill that must be developed. Moreover, I suggest that the physical variables of occupation surface slope and material deposit thickness can also profoundly affect GPR imagery of ephemeral archaeological deposits, which can be quite thin in relation to the sedimentary units in which they occur (Thompson 2011; Thompson and Storey 2010). There are a tremendous number of variables that can affect AGP applications.

VERBERIE LE BUISSON CAMPIN

The identification of subsurface archaeological material accumulations (in essence delimiting site and activity area boundaries) at archaeological sites is a fundamental purpose of ground-penetrating radar surveys in archaeology. This chapter describes the use of ground-penetrating radar to test the scale and configuration of unexcavated archaeological materials at the French Upper Paleolithic open-air site of Verberie le Buisson Campin. This involved the identification of and discrimination between high-amplitude radar reflections of both unexcavated anthropogenic material and previously excavated, disturbed sediment at an open-air Upper Paleolithic site. The main question to answer in this section is how one may discern between radar imagery (high-amplitude reflections) of both unexcavated (in situ archaeological) and previously excavated (disturbed and displaced sediment) materials at this site, in order to inform us about the use of space in unexcavated areas. This was necessary because at Verberie the reflections were found to be broadly similar in some cases. Detailed here is one example in which the use of high-resolution radar imaging and 3D rendering assisted in the identification of in situ archaeological deposits, and in discerning high-amplitude radar reflections from unexcavated archaeological accumulations and "false-positive" reflections caused by disturbed sediment from prior excavation activities, allowing reliable prospection and testing of archaeological material scale and configuration.

Although VBC has been extraordinarily productive and informative for analyses of Magdalenian subsistence organization, the excavated sample consists of less than perhaps 10 percent of the site surface area available for human exploitation during the Pleistocene. This presents a rather vexing interpretive problem since whereas VBC represents one of the largest Magdalenian archaeological datasets in the world (Bodu 2010; Enloe and Audouze 2010) conclusions based upon analyses of its content and material inventory derive from what is potentially a very incomplete sample (Thompson 2011). Moreover, if the site is in fact larger than previously known, then barring excavation of the entirety of the Pleistocene point bar on which it sits the archaeological inventory of VBC will also always be to some degree incomplete. Similarly, analyses of its material inventory and distribution (content and configuration) would also be incomplete. This is another very large and interesting problem since VBC represents a crucial dataset for understanding human adaptations in northern Europe during presumably large-scale environmental changes over the terminal Pleistocene. If we cannot directly sample archaeological content through excavation of the total area, could we at least estimate unexcavated archaeological configuration by some other means? Could we use ground-penetrating radar to do this at a site lacking monumental architecture, such as at an ephemeral Paleolithic open-air encampment?

A ground-penetrating radar (GPR) survey was undertaken at VBC in 2006, under the auspices of the Centre Nationale de Reserche Scientifique (CNRS) and the University of Iowa (Enloe et al. 2007). Although this survey was and is a very large undertaking, it was not a comprehensive, systematic survey of the entire point bar. Among several research agendas were prospection for unexcavated materials, basic feasibility-testing of ground-penetrating radar in a Paleolithic context as this study represented one of the first such attempts, and the development of methodologies for applying near-surface geophysics to prospection for very small-scale archaeological materials and features. The 2006 radar survey at VBC was not simply for identifying promising locations to excavate, although this basic prospecting was undertaken for ground-verification excavations conducted during the 2008 and 2009 seasons. An explicit aspect of the research design was to try to use ground-penetrating radar as a tool for the evaluation of archaeological material scale and configuration in order to address issues of demography and domesticity at VBC (Enloe 2010; Thompson 2011). As GPR has been used primarily at post-Neolithic sites with monumental architecture, any application at a Paleolithic site offers potential for methodological innovation.

The rationale was to attempt the location of possible unexcavated stone-lined Magdalenian hearths and associated artifacts near an excavated and retired area (excavated down to sterile sediment) that has produced numerous artifacts as well as at least one hearth (Audouze 2010; Enloe et al. 2007). The topic of this study involves one aspect of that larger methodology, concerned with establishing reliable means of distinguishing between radar reflections of nonarchaeological material and reflections of actual, unexcavated, and legitimate in situ deposits known to exist at this site, with direct testing through excavation in 2008 and 2009. Prior to the performance of possible future comprehensive radar surveys at VBC, a reliable method for separating reflections of archaeological items from those of other materials was first needed, in order to demonstrate the presence of intact archaeological material.

In several areas reflections of subsurface materials were located that seemed possibly archaeological but which under more intensive scrutiny proved to be what are herein termed "false-positives," or suggestive but misleading radar reflections of subsurface sediment disturbances formed during excavation and backfilling episodes at the site prior to the radar survey of 2006. These disturbed sediment materials produced high-amplitude reflections that were superficially similar to those of in situ archaeological deposits at VBC. Focusing precious excavation resources on such areas would have proven both fruitless and wasteful since they had already been excavated, the artifacts removed, and then backfilled. This section of the volume describes a process of practical, problem-solving methodology for distinguishing between reflections at this site, with potential application at other sites with known or inferred unexcavated portions containing archaeological materials, perhaps also relevant to other geophysical methods.

Most individual archaeological material features at VBC have dimensions of submeter size, buried at elevations of between −1.1 m and −1.4 m below the contemporary ground surface in a sediment matrix of fine sandy silt, with the highest densities of artifacts occurring between −1.27 and −1.37 m (Audouze and Enloe 1997:199–200, Figures 1 and 2). Outlier artifacts have, however, been located at VBC at depths ranging between a minimum of about −70 cm and a maximum of −1.5 m (Audouze and Enloe 1997). In many cases such materials and features are concentrated in larger accumulations or middens extending over areas of several square meters. Potential archaeological material was identified according to high-amplitude reflection depth, scale, configuration, extent, and shape. Such archaeological materials consist at VBC of flint tools, lithic knapping debris, bones, and fire-cracked limestone arranged in variable thicknesses and frequencies, mostly having elongated, amorphous, or rounded shapes that appear to lack cardinal directional or rectilinear orientations (Audouze and Enloe 1997). This is true both of the actual artifact accumulations and sterile, empty spaces between accumulations. The archaeological strata are overlain by an agricultural overburden of approximately 50–60 cm in depth, composed of a very fine silty loam situated atop the Pleistocene matrix sands. This overburden is an active agricultural plow zone.

Previous excavations at VBC have resulted in the location and removal of lithics, bones, and fire-cracked rocks from the sand matrix from exposures such as *sondages*, or test trenches, and large rectangular *décapages* (wide areal exposures of tens of m2), and test pits with squared or rectilinear shapes and standardized sizes of sensible metric dimensions (such as 1 × 1 m^2 units, trenches of 1 or 2 m width and variable lengths, and *décapages* of 5 × 10 m^2, etc.). At the cessation of excavation seasons, exposures were backfilled with excavated sediments. Efforts were taken to keep different sediment types separate (loam topsoil in one pile, archaeological matrix sands in another pile) in order to replace sediments in the opposite order from which they were excavated. This allowed for a general reconstruction of the actual graded sediment column, with the silt loam left on top in order to preserve the agricultural productivity of the soil. Despite the care taken in backfilling, the images presented in this short report suggest that the radar reflections of sediment disturbances caused by previous excavation exposures and backfills can in some cases mimic reflections from in situ and unexcavated archaeological materials. Radar waves reflect off changes in material density and electrical conductivity, which are in turn based upon variation in moisture content (Conyers 2013). Because previous excavations since 1975 have resulted in sediment disruptions to the matrix at VBC, these disruptions (with slight differences in material density and moisture content) produce reflections of radar waves. Distinguishing between reflections of actual archaeological material and disturbed sediment from previous excavations thereby became extremely important at VBC.

REFLECTIONS FROM INTACT UNEXCAVATED ARCHAEOLOGICAL MATERIALS

VBC was excavated annually between 1975 and 2002 (Enloe 2010). The last year of extensive excavations at VBC (2002) resulted in the exposure of a three-meter north profile wall revealing the extension of intact archaeological materials northward. In 2006, a team from the University of Iowa and the CNRS performed a GPR survey of the site. The site area was surveyed using the University of Iowa's GPR panoply of a Geophysical Survey Systems, Inc. (GSSI) SIR-3000 radar unit, with 400 MHz and 900 MHz antennae. Sample (profile) intervals were scaled between units of 10 cm and 25 cm for the 900 MHz antenna, 25 cm for both antennae, and 25 cm, 50 cm, and 100 cm for the 400 MHz antenna (Enloe et al. 2007). Standard radar settings were used for passes: 512 samples per scan, 64 scans per second, 40 scans per meter, hi-pass filtering at 800 MHz and low-pass filtering at 200 MHz. Metric distance data was entered automatically by a survey wheel, with ticks at 1m intervals. The velocity of the radar wave through the sediment at VBC averaged 0.1058 m/ns. This allowed for relatively accurate estimations of time-depth by simple multiplication of the wave velocity (0.1058 m/ns) with various time-depth measures in nanoseconds (for example, time-depth of 6 ns = 6 ns* 0.1058 m/ns velocity = 0.6348m, for a subsurface depth of about –0.63 cm). Time-slice maps in this report were generated at one-half nanosecond depths, which given the radar velocity of 0.1058 m/ns amounts to intervals of approximately 5.3 centimeters in depth (Storey personal communication). Our radar survey successfully relocated intact archaeological material exposed in a wall section in 2002 at the season's end.

Given the exceptionally fine spatial control afforded by the use of the theodolite grid in locating this GPR grid, we can be relatively certain the material we saw in profile (fire-cracked limestone cobbles and flint lithics seen in exposure) corresponds to the other reflections. Since the time-depth of the materials noted in the radar profile are about 12 ns the relative depth according to the wave velocity of 0.1058 m/ns would be approximately –1.27 m, which corresponds to the depth assigned to the highest frequency of previously excavated archaeological material at this site (Audouze and Enloe 1997:200). The truncated (squared-off excavated) southern margins of the actual stone and flint accumulations appeared as truncated southern margins of the radar reflections. The depth, scale, and configuration of the reflections thus supported their interpretation as the same intact archaeological deposits visible in 2002. It is, however, interesting to note that a group of flint artifacts only barely showed in the radar profile and was essentially invisible in the time slices, whereas the limestone cobbles were highly reflective and visible in all imagery of this grid.

Directly west of the material above was located another series of high-amplitude reflections from subsurface material deposits, albeit from a region unexcavated at the time of the GPR survey in 2006. These reflections indicated the presence of a roughly semicircular deposit, arcing to the northwest, with a concavity oriented toward the southwest, similar to a classic Binfordian forward toss-zone (Binford 1980) approximately 7 m in length and of variable width, at between 7 ns and 12 ns time-depths (or between -.74 m and -1.27 m subsurface), well within the range of known artifact depths at the site. The area was excavated in 2009, with the areas chosen for excavation focus in 2009 being premised upon image analyses of the 2006 radar data. Four units were excavated directly in the semicircular reflections, producing many flint tools, lithic debris, and bones, with one of the multiple superimposed levels shown in the photographs collected in the field (Audouze and Enloe 2010). In the case of this semicircular deposit, which was tested directly through excavation in 2009, the choice to dig in the units indicated by high-amplitude reflections was extremely productive. These semicircular shapes will assume much more importance below.

SPURIOUS REFLECTIONS FROM DISTURBED, PREVIOUSLY EXCAVATED SEDIMENT

Other areas of the site also produced high-amplitude reflections suggestive of intact, subsurface archaeological materials that were visually similar to the authentic ones which partially confounded the selection of areas for excavating in 2009. The velocity of the radar wave through the sediment in a different grid was also 0.1058 m/ns. The high amplitude reflections indicated materials at time-depths ranging between 9 ns and 15 ns, or between -0.95 m and -1.59 m, at a much deeper maximal depth than the average depth range of artifacts found at VBC. The area was surveyed at a maximal time-depth of 15 ns, so nothing below that time-depth was surveyed for this area. The reflections fall partially within the -1.27 m to -1.37 m mean depth range of maximum artifact concentration, thus the depth analysis was not conclusive.

Three areas indicated high-amplitude reflections; one, a group of rounded reflections to the left or grid west of the area was known by the grid placement to correspond to a buried concrete datum used for calibrating the theodolite, thus representing reflections from an area disturbed over multiple, successive excavation episodes. The other two were less securely known from previous photos and artifact maps, and consisted of a central, "linear" feature extending from the grid north to the grid south of the area and a third group of reflections to the extreme grid east of the area. In each of the latter two cases, the individual high-amplitude reflections were roundish or roughly circular, which as above was a criterion for cursorily identifying archaeological materials. Each group of reflections also seemed to feature different aggregate shapes, although only the reflections of the feature in the center of the grid seemed to be symmetrical. Finally, individual reflections in these two groups appeared to be generally of submeter size, which was another criterion for identifying reflections of archaeological materials at VBC.

Since the depth analysis of reflections was by itself inconclusive, alternate means were necessary to rule out potentially unproductive areas for excavation during the 2009 season. Whereas archaeological interpretations of GPR imagery are based in large measure on pattern recognition, patterns can obviously be deceptive. By definition, intact subsurface material has not been previously excavated, whereas material disturbed from prior excavations has. In this way, a crucial methodological and interpretive problem at VBC for the focal areas of the 2009 excavation season involved the discrimination between patterning of high-amplitude reflections of actual in situ archaeological deposits and what are herein termed "false positives" or patterning of high-amplitude reflections that are suggestive of apparent in situ deposits but that represent prior excavations and lack current archaeological import. By what means can the reflections of actual and apparent archaeological materials be segregated? How can we tell the true from the false? Solutions to such interpretive difficulties can come from one of GPR's most obvious selling points: its capacity to produce images indicative of three-dimensional subsurface patterning (Conyers 2004; Conyers and Cameron 1998; Leckebusch and Peikert 2001). Images generated from radar data can also be combined with other linear, directional, and scalar devices that can serve to clarify interpretations of subsurface material patterning.

FALSIFYING THE FALSE POSITIVES AT VERBERIE

It seemed intuitive that the key to differentiating between potentially productive and unproductive areas for 2009 excavations at VBC would by necessity be found in analyses of radar imagery. For this site, only the 2006 GPR data exist and no further geophysical surveys have yet been performed at it. A simple question was then posed: How would false positive reflections of sediment disturbances, such as trenches, test units, and wider *décapage* exposures from prior excavations appear in the radar imagery from VBC and what things would clearly indicate their falsity vis-à-vis reflections of in situ subsurface archaeological material? It was reasoned that reflections of actual archaeological material might perhaps portray orientations distinct from more modern excavation disturbances since it could be assumed that the Magdalenians did not construct their sites according to contemporary archaeological spatial norms of directionality and linearity, especially in the use of gridded planar space. It was therefore reasoned that if high-amplitude reflections in the radar images from VBC showed evidence of strong directionality and linearity oriented to the VBC grid coordinate plane (again, which is slightly offset from true magnetic north, which is NE of VBC grid north) then such reflections could be assumed to represent sediments reworked and disturbed during previous archaeological excavations.

For example, the imposition of grid coordinate planes and other spatial references during excavations is used to provide modern researchers reference points. Such referents, however, have no real bearing on the spatial material distributions past humans created. There is therefore no reason to expect our grids and other spatial references (directional markers, various metric devices indicative of scale, etc.) to correspond other than cursorily to past material orientations. Yet, given the dynamics and methods of contemporary excavation, we can expect that recent excavations (in trenching, wider exposures, and test pits) at sites might feature similar orientations, directionality, and linearity since we all generally use the same spatial references, directionality, and scalar dimensions today. At VBC, evidence for linearity and directionality in subsurface materials can be regarded as a proximate index of previous excavations, regardless of material depth, and whether in plan-view slices or vertical radar profiles. Unusual time-depths of reflections and of radar profile hyperbolas might also be expected to occur where excavators systematically proceeded *below* artifact-bearing levels down to sterile sediments, leaving disturbed sediment in backfill. In fact, at VBC, excavations were routinely extended to sterile depths below all artifacts so as to retire areas of the site.

A series of time slices each 1 ns in thickness (9ns through 14 ns time depth) was combined into a three-dimensional model of isosurfaces, indicating that what appeared to be a group of circular features actually presented dimensions highly suggestive of a one-meter-wide test or drainage trench, strongly oriented to grid north-south according to the VBC grid coordinate plane. Another mechanism to gauge linearity and directionality involves the superimposition of a grid system atop GPR imagery to simulate the presence of the established site spatial coordinate plane. Some of the amplitude maps of this area of VBC showed reflections indicative of very strong linearity and directionality, especially in relation to the virtual grid used by the laser theodolite. In other words, the radar reflections seemed virtually to correspond to the gridlines. In this example, practically all visible features were seen to correspond to or fall within gridlines or inside actual grid units of the simulated

VBC grid system used for excavations. It is highly unlikely that legitimate archaeological deposits would conform so closely and consistently with a contemporary spatial referent, as, for example, the superimposed grid.

One of these directional features was a trench 1 m in width. This trench was visible from 9 ns to 15 ns, as was another larger directional "decapage" excavation feature located east of the trench. It appears that the active plow zone has obliterated the upper levels of both features at about the same elevation. It also appears both the apparent trench and excavation exposure were excavated below artifact levels to sterile sediment, both to at least 15 ns (= –1.59 m). It is interesting that the two features would both be truncated at the tops and yet would also terminate at the same time-depths (i.e., 9 ns and 15 ns). The use of 3D imagery in accompaniment with directional and linear references simply clarified the impression of these features as disturbances from prior excavations (Thompson 2011; 2014a). It appears that the prior excavation exposure, located to the right of the time slices, has strong three-dimensional depth, linearity, and directionality as well. Features with both two- and three-dimensional correspondences with the VBC grid coordinate plane, in other words, linear width, depth, and breadth, along with grid-oriented directionality, can be safely assumed to relate to prior excavations. In this case, the shape and scale of the feature (~5 m width, uncertain length) is suggestive of rectangular previous *décapage* excavation exposures, which involve maximal lateral exposures of presumed occupation surfaces across space. The author personally participated at this site over four field seasons and was involved in the excavation and backfilling of two such rectangular exposures, consisting of 5 m × 10 m pits, oriented to grid north.

The association of linearity and grid-oriented directionality identifying previous excavation disturbances at VBC suggests that such properties could be used in prospection at other such sites, so long as at least some of both excavated and unexcavated material is present and that such previous excavations were undertaken according to contemporary norms of spatial provenance and control. If prior excavations at a site have revealed repetitive instances of nonlinear, non-directional artifact patterning in the in situ materials, lacking grid-oriented linearity or directionality, the location of strongly linear or directional "false-positive" reflections (or various anomalies in other geophysical methods) could indicate prior excavation features as opposed to archaeological materials. This would be especially valuable if available photos or maps of excavations did not preserve locations of all prior exposures, or if such graphic materials were unavailable.

The 2006 VBC radar survey indicated several areas that seemed to indicate presences of artifacts but that under further scrutiny have proven to be highly suggestive of subsurface disturbances caused by earlier excavations. Given the labor- and capital-intensive nature of excavating, especially internationally, this brief report suggests that using GPR's imaging capabilities, along with standardized spatial references, to assess the shape, configuration, and orientation of reflections from buried materials could assist in discriminating between legitimate and spurious archaeological deposits. In similar applications at other sites, even with other geophysical methods, the key would be to establish a means of correlating excavated material with known or suspected unexcavated portions through spatial means (i.e., do reflections of unexcavated material correspond with known excavations in photos or maps?), and identification of suggestive directionality, linearity, or concordant depths in radar reflections, magnetic or geoelectric/resistivity anomalies.

The archaeological significance of this methodological overview consists in the ability to discern in some cases between actual archaeological material and other materials that mimic it in radar reflections (backfill disturbances, test pits, trenches, etc.), as well as a demonstration of GPR's utility in application to Paleolithic archaeology. This is one of the first such attempts at a Paleolithic site. Anthropological significance is found in the ongoing development of geophysical means to assess small-scale, ephemeral human behavior and spatial organization. Generally, geophysically visible cultural phenomena have, to now, focused on larger scale material that is easily visible, often at Neolithic or later permanently settled sites (with walls, pyramids, temples, etc.). This has tended to deemphasize geophysical analysis of smaller scale, ephemeral Mesolithic and Paleolithic sites dating to earlier periods. The boundary has been therefore partially methodological (i.e., how *can* we investigate Paleolithic open air ephemeral sites geophysically?), partially temporal (geophysics at Neolithic and later sites on one hand, lack of such investigations at earlier sites on the other), and partially structural (permanent villages, towns, fortresses, castles, cities vs. impermanent temporary encampments).

The examples above suggest that we are beginning geophysically to add some key dimensions to the Upper Paleolithic as well through radar surveys. The promising results in this report may apply also to other geophysical techniques, allowing for much fuller archaeological analyses. Although the 2006 radar survey at VBC did not sample the entirety of the Pleistocene point bar, it has demonstrated the utility of using near-surface geophysical survey as a means of addressing issues of Paleolithic archaeological material scale and configuration. Deposits of intact unexcavated archaeological material were provisionally identified in the 2006 survey data, and were directly tested substantively in 2008 and 2009. The 2009 VBC excavations were particularly productive. For the present report, the results can be interpreted as a successful demonstration of the methodology at this site for locating intact archaeological deposits, with possible applications using other geophysical methods at other sites.

TOWARD "SEEING" MATERIAL SIGNATURES OF HUMAN BEHAVIOR AT VERBERIE

In terms of using radar survey to address the more high-order theoretical demographic/domestic issues at VBC, in order to make any conclusive statements we would first need to conduct a comprehensive radar survey of the entire Pleistocene point bar on which the site is located, if not the entire agricultural field. Such a survey should indicate the existence or absence of other possible household/hearth areas at VBC. Based on the results of the 2006 radar survey, it is possible to say that unexcavated materials do remain at VBC. Yet, since the 2006 survey was not comprehensive, we have no present means to say just how extensive these archaeological materials may be, nor are we in a position to describe their actual scale and configuration at the level of the entire site. Given the decades of research at this outstanding site, which represents a tremendous collective archaeological investment in terms of money, time, and effort by dozens of scholars, a comprehensive radar survey of the entire site certainly seems warranted.

Yet, the ground-penetrating radar data that were collected at VBC in 2006 are extremely valuable in their own right, in that they appear to indicate some legitimate signatures of human behavior, far beyond site prospection or ascertaining the site's dimensions. As this material is introduced, one can also gain examples for how to construct linking arguments for connecting AGP data to human behavioral indices, through methodological integration, in the present case with detailed faunal analysis of reindeer (*Rangifer tarandus*) bones collected over nearly three decades from VBC. As was briefly introduced above, the faunal analysis resulted in the discovery of distinct patterns of bone conditioning from different occupations of the site, each with a particular shape, scale, configuration, orientation, bone frequency, and distribution. These bone conditioning patterns also featured distinct shapes: one circular and "concentrated" or compressed, the other semicircular and diffuse. In one's opinion these spatial configurations (shapes) would be visible to ground-penetrating radar if not other AGP techniques. These differential bone conditioning patterns with different shapes and configurations also produced highly variant mortality profiles once dental attrition analysis was conducted to access the population structure of the animals present in this data set.

SITE LAYOUT AT VERBERIE

The site of Verberie le Buisson Campin (VBC) is located within the Paris Basin (PB), a large sedimentary basin in the northeast of France, in which the city of Paris is located in the center (Audouze 1987:183–184). The Paris Basin is surrounded and defined by uplands; the Ardennes to the northeast, the Vosges to the east, the Morvan to the southwest, and the Massif Central to the south (Audouze 1987:183). The small town of Verberie (for which the site was named) is located near the confluence of the Oise and Aisne rivers at Compiegne. The Paris Basin itself formed during Permo-Triassic times, as a topographic low near the Tethys Sea, and features three primary lithologies: Lower Triassic sandstones, Middle Triassic chalks, and Upper Triassic (Keuper) limestones (Bourquin et al. 1997:670). The carbonate sedimentary rocks of the Middle and Upper Triassic formation in the PB contain deposits of high-quality flint, mainly in the valleys carved by fluvial incision into the bedrocks. The contemporary regional dendritic river drainage system is dominated by the River Seine, flowing east to west (Audouze 1987).

Approximately after the Younger Dryas (12.5 ky to 11.8 ky) the Oise River underwent a transition from a braided to a meandering stream channel, accompanied by decreases in both sediment and water discharge, and channel incision (Pastre et al. 2002). As the earlier Pleniglacial riverine deposits (coarser sands and gravels) were supplied by cryoclasty and intense sedimentation, the transition in channel morphology probably marks the lapse of glacial conditions and the onset of Holocene climates (Pastre et al. 2002). The VBC archaeological deposits occur within a matrix of homogenized, extremely fine calcareous sandy silt deposited during a general aggradational phase that occurred across many river channels in northern France, marking transition to Holocene climatic conditions (Cordier et al. 2005; Pastre et al. 2002). Since VBC archaeological materials do not occur within the coarse cryoclastic gravels and sands underlying the fine silts, it's apparent the site was occupied during milder climatic conditions. The change in the river's hydraulic geometry, decreased sediment discharge, and the occurrence of archaeological occupations probably also coincides with the return of browse vegetation to the area, consolidating slopes and fortuitously retarding erosion of the site.

The human occupations at VBC are located on the south (cut) bank of the Oise River, a tributary of the Seine, upon a Pleistocene terrace remnant, northeast of the town of Verberie, west of the town La Croix-St. Ouen. The site is situated within a contemporary agricultural field, the planting of which is done on a crop rotation basis. Figure 6.1 below provides a schematic plan view of the site, while Figure 6.2 below highlights the different locations (secteurs) at VBC from which the two distinct bone patterns were excavated. It is pertinent to note that on both the following figures the units are drawn at a 20 m^2 scale.

During the field season of 2006, a GSSI SIR-3000 ground-penetrating radar unit was used at VBC. The sampling format included GPR samples taken at one-meter, half-meter (50 cm), 25 cm, and 10 cm sampling intervals along both x and y axes, over excavated and unexcavated portions alike, using both 400 MHz and 900 MHz antennae.

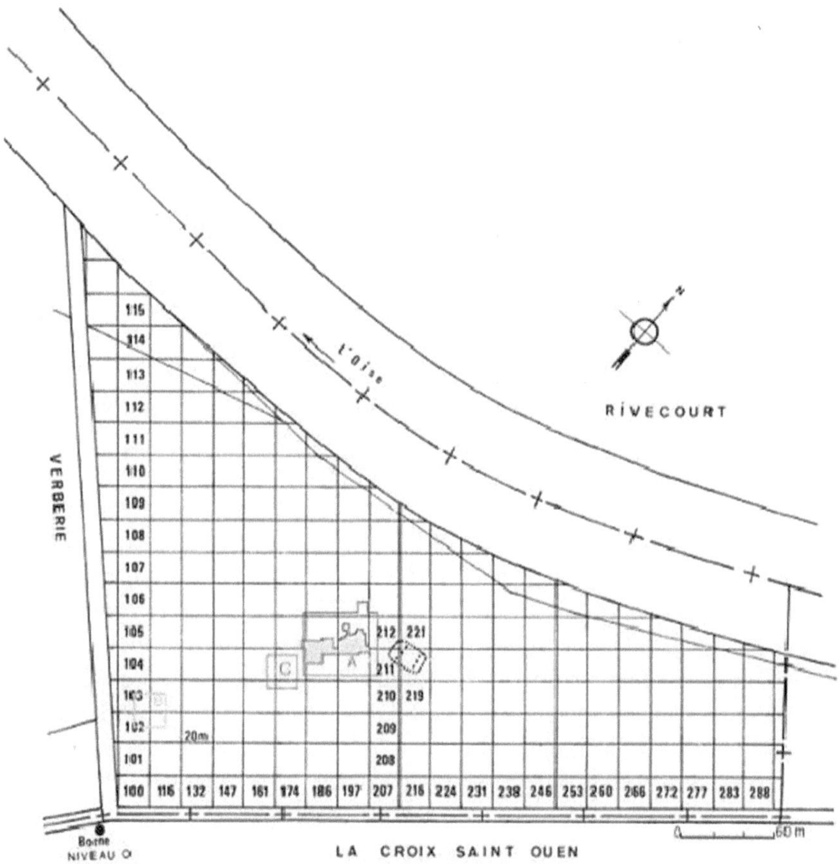

Figure 6.1. Site plan of VBC. 2006 radar grids located inside the polygons at center. Each square unit on site plan is 20 m²

When promising subsurface targets were located, finer sampling formats (i.e., decimeter-spaced transects and tie lines) were used in order to gain higher subsurface resolution for more accurate imaging and 3D modeling. GPR "time slicing" of any detected target surfaces was then compared/contrasted with vertical provenance data from previous excavations in order to augment our understanding of both horizontal and vertical components of unexcavated in situ archaeological deposits. The sampling interval was adjusted based upon field requirements; i.e., some features were detected at lower resolution, to be sampled later at a finer scale. One explicit aspect of this study was methodological development and to determine the feasibility of applying high-resolution GPR subsurface sampling at an open-air Paleolithic site with small-scale features.

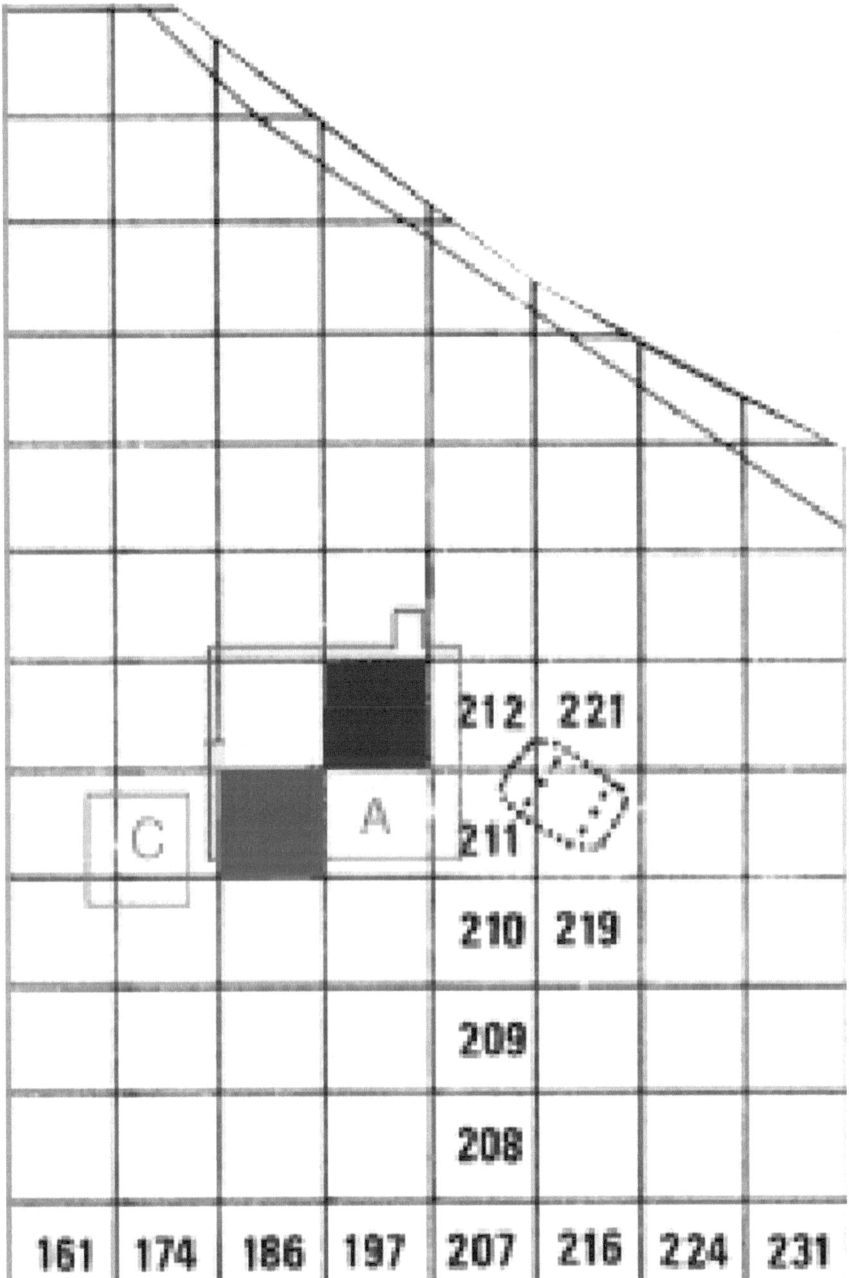

Figure 6.2. Zoom-in view of VBC site plan showing the individual secteurs 190 and 202 (light and dark squares) from which the two distinct patterns were ascertained.

Processing of the GPR data was accomplished primarily through the use of a software suite and methodological approach that could be called the Conyers-Lucius Method.[1] Basic GPR plan-view mapping was done in Golden Software's Surfer 9. Profile views of individual GPR sections were constructed using Conyers-Lucius GPR Viewer.[2] Alignment and slicing of the GPR sections was accomplished using Lucius-Conyers GPR Process.

FAUNAL METHODOLOGY

The faunal methodology for this study featured a detailed analysis 1) to assess taphonomic agencies, 2) to test the visibility and configuration of behaviorally structured patterning of bone condition, spatial patterning, and disposition, and 3) to gauge faunal evidence for systematization, standardization, and consistency in bone processing behaviors at VBC. This faunal assemblage consists of material from two separate areas of VBC: one excavated from secteur 190, the other from secteur 202, located as shown on the figures above. One potential caveat was that differential excavation methods were used for the 190 and II.22 faunal assemblages, since these data were collected during different decades (190 was excavated in the mid-1970s, 202 II.22 from the late 1990s and 2000s). Whereas each secteur at VBC was excavated in 1 × 1 m units, each artifact from II.22 was excavated with precise, three-dimensional coordinate provenance data provided by a total station. The photocopied dossiers that I have for VBC 190 (field seasons 1975, 1976, and 1977) do not record individual artifact elevations within 1 × 1s, instead listing the artifacts to manual piece-plotted X-Y provenance only (no 3D provenance, only 2D). There should be no persistent epistemological concerns regarding the lack of elevation data for VBC 190 fauna, since this material derives from the same site and is limited to analysis of *Rangifer* specimens only and furthermore since this analysis is concerned only with horizontal bone distributions and not depth.

There is an additional methodological issue relating to the international location of portions of both the VBC 190 and II.22 collections (some of the material is located in Iowa City, Iowa, some in France). Although every effort has been made to study this material in toto, some isolated VBC 190 specimens were spotted and identified at a regional museum in France. These items have been included in faunal quantifications. However, two mandibular fragments were also located in a Compiégne museum in July 2009 that unfortunately contained excellent articulated teeth. These teeth could have been used in constructing age profiles for VBC 190. So those minor elements are missing from this analysis as well as the earlier works from which it was drawn (Thompson 2011, 2014). In one's opinion, the teeth at the museum would have added to an age-skew in one of the mortality profiles. Such are research complications to which we must often simply accommodate ourselves.

Moreover, in spite of two trips made to France to analyze this fauna during the field season of 2009, negligible additional specimens (fewer than five) from II.22 were located too late for incorporation into this volume. This research is therefore contingent upon slight future revision should those specimens from both secteurs ever be analyzed in the future. Microsoft Excel and dbase were the primary data handling software for recording artifact attributes. SPSS was used for performing statistical analyses, while Golden Software Surfer 9.0 was used to make spatial maps of bone density-distribution, calibrated at the level of individual skeletal elements (i.e., rib, femur, metacarpal, etc.). Although faunal refits between paired elements were sought, no interassemblage pairings were found, between 190 and II.22. Since it is often possible more accurately to gain prey species demographic information from dentition, the dental sample was analyzed first, with dental identifications conducted as follows: position in the dental arcade, taxon, and side, maxillary vs. mandibular dentition, sex, and age as possible, using crown height methods for aging. Postcranial elements were identified to taxon, element, side, sex, and age (Thompson 2011, 2014). The postcrania were also quantified for estimates of NISP and MNI. The teeth were also quantified for estimates of NISP and MNI, a process to serve as an internal check on similar estimates drawn from the skeletal sample (Thompson 2011, 2014a). As the current volume is dedicated to archaeological geophysics, the above should suffice as a description of the faunal data linked methodologically to AGP data.

The salient results of the faunal analysis are as follows. The skeletal NISP for secteur 190 was 1,566 drawn from a total sample population of 1,855 bones, teeth, and fragments; The skeletal NISP for secteur 202 II.22 was 386, drawn from a total sample population of 1,610 bones, teeth, and fragments (Thompson 2011, 2014a). For this faunal assemblage the dental MNI was regarded as the most accurate estimate of individual reindeer present in the sample from each secteur. The dental MNI for secteur 190 was 12, while for secteur 202 II.22 the dental MNI was 7. For secteur 190, the mortality profile indicates an extremely old mean individual age per animal at 12.99 years, with multiple individuals over 17 years as estimated by crown height attrition; the mean individual age for secteur 202 II.22 was estimated at 5.54 years, with a maximum of 15 years for one individual. The associated postcranial material from 190 was also in much better condition and preservation than that from secteur 202 II.22, consisting in many cases of entire or nearly complete bones with little fragmentation, which can be seen by comparing the dental MNIs from each secteur with the associated skeletal NISPs. The dental MNI from secteur 190 was for 12 individuals in comparison to an element NISP of 1,566; the dental MNI from secteur 202 II.22 was for 7 individuals in comparison to an element NISP of 386 (Thompson

2011, 2014). The ratio of dental individuals to identifiable fauna specimens for each secteur was therefore 1:130.5 for secteur 190 and 1:55.14 for secteur 202 II.22. The ratio of individuals to skeletal NISPs is an index for how much more bone-processing and transport off-site of postcranial elements occurred in the secteur 202 II.22 occupations in relation than those in secteur 190, for which most postcrania were simply disposed of on-site.

Most of the faunal material from VBC 190 is in unusually excellent condition, in many cases whole bones or minimal evidence of marrow-cracking and little to no bone erosion, as indicated above. The skeleta from VBC 202 II.22 was much more fragmented, profoundly more so, showing extensive signs of anthropogenic processing which has been well-documented previously at this site (Enloe 2004, 2006). While space precludes a comprehensive examination of taphonomic conditioning present in this faunal assemblage, fuller treatment of such factors is present in Thompson (2011, 2014a). In general, the taphonomic effects visible on the fauna from each secteur were highly comparable. There are few to no indices of any sort of fluvial or other mechanical sorting or redeposition of the bones (Enloe 2004, 2006, Thompson 2011, 2014a). The sum of all lines of evidence indicate that anthropogenic patterning is the primary causal mechanism of the conditioning, spatial position, configuration, orientation, and distribution of faunal material at VBC, and this is true for all heretofore excavated areas of the site (Enloe 2004, 2006; Thompson 2011, 2014a). To summarize the two distinct patterns, we can observe a circular, 2 m^2 dense concentration of whole or minimally broken and processed reindeer bones, fire-cracked limestones and sandstones, and lithics in VBC secteur 190, containing a minimum number of 12 individual animals with a mean carcass age of 12.99 years and maximum ages of over 17 years for multiple carcasses. In VBC secteur 202 II.22, we observe a much more diffuse, semicircular distribution of highly fragmentary and heavily processed reindeer bone (presumably from marrow-cracking and subsequent grease-boiling), containing a minimum number of 7 individual animals with a mean carcass age of 5.54 years and a maximum age of 15 for one individual carcass.

Given the extensive previous research conducted on the fauna and geomorphology at VBC, we can conclusively state that the primary conditioning mechanism on material patterning at the site was human behavior (Audouze and Enloe 1997; Enloe 2004, 2006; Thompson 2011, 2014). All data available at present suggest that the main difference between the secteur 190 and 202 II.22 fauna relate to differential economic processing of the two assemblages. It appears that the secteur 202 II.22 fauna was subjected to a much higher intensity of human modification through various processing activities, such as marrow-cracking of long bones, and grease-boiling of diaphyseal fragments and other elements such as flat bones and fragments thereof, distal articulations and epiphyses. Perhaps consequent to the sequential nature of such processing activities, the secteur 202 II.22 fauna were distributed more widely across wider areas, in semicircular configurations, with very few joint or vertebral articulations. Contrarily, the secteur 190 fauna indicate a much lower intensity of human modification. Whereas most of the secteur 190 long bones exhibited evidence of marrow-cracking, there were many whole long bones present, along with many articulated joints and vertebra, and virtually no evidence of the sort of smashing necessary for grease-boiling. For example, the respective ratios between dental MNIs and skeletal NISPs above (1:130.5 for secteur 190. 1:55.14 for secteur 202 II.22) reveal that any given specimen in secteur 202 II.22 was 57.75 percent less likely to be identifiable than any from secteur 190. Nearly all of the bones from secteur 202 were heavily fragmented. Why would the much younger reindeer carcass cohort from secteur 202 II.22 reveal so much more processing than the much older carcass cohort from secteur 190?

A number of anthropogenic agencies were considered to account for the differential spatial and behavioral patterning in the fauna from VBC 190 and 202 II.22, ranging from diachronic variation in hunting ability, to vagaries of stochastic herd compositions, to freak accidents of nature. Yet, none of the preceding would account for the differential *spatial* configuration (i.e., shape, distribution, and frequency of faunal specimens) observed between the two secteurs. An explanation for the different patterns has to account for both the variation in spatial configuration of materials present as well as the differential bone modifications. A seemingly good functional analogy for the secteur 190 pattern is a circular trash pit or dump, while the 202 II.22 pattern appears analogous to semicircular patterns described by Binford (1980) in his Nuniamut ethnoarchaeology. Why would lightly processed, geriatric reindeer (including articulated joints and vertebral sections) be dumped indelicately in tandem with lithics, cores, and rocks in a dense, circular trash dump in contrast to much younger fauna being heavily processed and dispersed diffusely in semicircular arrangements?

The much older-skewed mortality from VBC secteur 190 gives us an important clue when we consider economic utility, specifically fat content. An older reindeer cohort would presumably feature lower fat content and would exhibit more chronic degenerative conditions, such as tooth wear, osteoarthritis, and generally poorer health than a younger cohort. From that perspective, we can infer that the fauna from VBC secteur 190 were nutritionally simply not worth the labor investment in extensive economic processing we find in secteur 202 II.22. Variable fat content could therefore explain the different degree of bone processing between the two secteurs. In a relatively xeric and cold environment, such as Pleistocene northern France was, animal fats constitute one of the single most valuable resources, especially where little to no plant fats were available (Speth and Spielman 1983). Fat's nutritional value is due to its necessity in protein metabolism: without adequate dietary fat, proteins are not metabolized, which can lead to protein poisoning (Speth and Spielman 1983). What would affect the fat content beyond an animal's chronological age? Nutritional stress, fat depletion, and lack of adequate food sources could coincide, such as during winters, again in a harsh, xeric environment. Widespread availability of food, such as during springs and summers, could obviously alleviate fat-depletion. Seasonality thereby emerges as a potential conditioning agent for variable prey valuation and differential investment in faunal processing. A question to ask here would be whether seasonality has any potential *spatial* ramifications at Verberie?

DIFFERENTIAL SPATIAL DISPOSITION OF FAUNA AT VERBERIE?

To answer the question of whether seasonality has spatial ramifications at VBC, we have several lines of evidence available beyond faunal processing intensity.

The spatial configuration, orientation, and distribution of processed fauna certainly bear some consideration. The shape of material deposits from each of the secteurs has been discussed and we will now investigate these features in graphic detail below. Figures 6.3 (metacarpal), 6.4 (femora), and 6.5 (ribs) below show the spatial scales and configurations of fauna from secteurs 190 and 202 II.22. These figures are faunal density-distribution maps created in Golden Software's Surfer 9.0. Using Surfer 9.0, digital data grids were created for both of the 20 m^2 secteurs, VBC 190 and VBC 202 II.22, each consisting of twenty

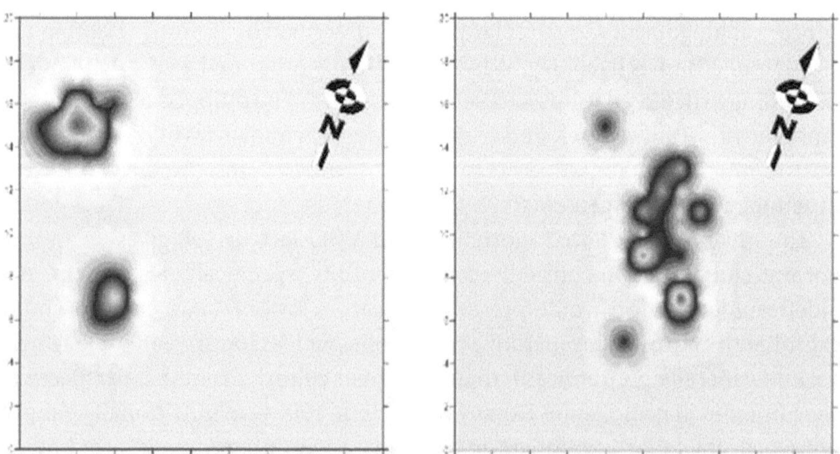

Figure 6.3. Faunal density-distribution of MC at VBC. Secteur 190 left, Secteur 202 II.22 right. Note the very different patterning and the large circular feature in 190 left and the semicircular feature in 202 right.

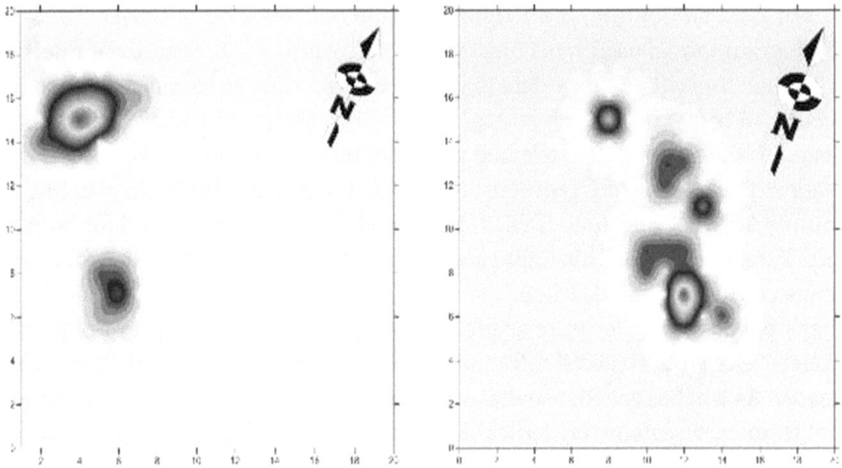

Figure 6.4. Faunal density-distribution of femora at VBC. Secteur 190 left, Secteur 202 II.22 right. Note the very different patterning and the large circular feature in 190 left and the semicircular features in 202 right.

1 m² units. These 1 m² units were further divided into four 0.25 m² quadrants. In other words, the following three figures portray the faunal frequency per 1 × 1 m unit accurate to within 25 cm. The 25 cm interval was chosen due to data discrepancies between the 190 and 202 II.22 provenance data. The 190 piece-plotting provenance data available were only accurate to this interval, so to make these density-distribution maps comparable on an intra-site basis a lowest common metric was selected.

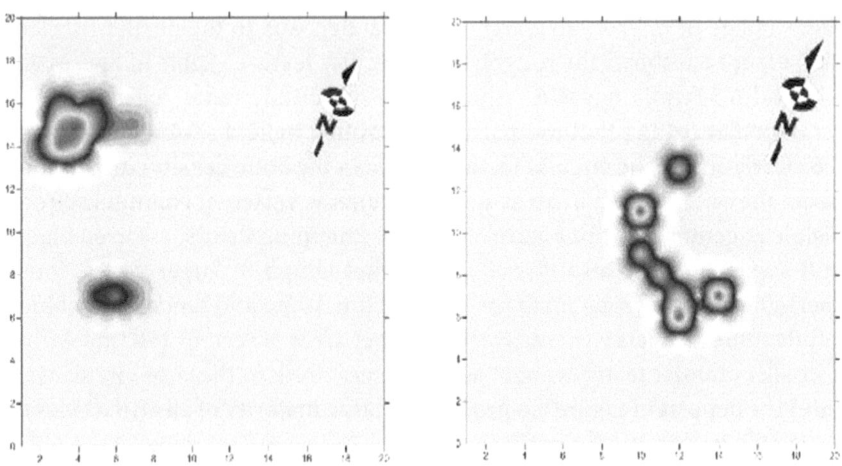

Figure 6.5. Faunal density-distribution of ribs at VBC. Secteur 190 left, Secteur 202 II.22 right. Note the very different patterning and the large circular feature in 190 left and the semicircular feature in 202 right.

For data smoothing, the Kriging method was selected in Surfer 9.0. The Variogram model used was Linear, Slope = 1, Aniso = 1,0. Search parameters were 4 sectors with 64 max data to use from all sectors, 16 max data from each sector, 16 min data in each sector, and 3 blank nodes for empty sectors. The Search Ellipse parameters selected were Radius 1 = 7.0, Radius 2 = 7.0, and the angle = 0. The grid line geometry was set as follows: X Miminum = 0, Maximum = 20; Y Minimum = 0, Maximum = 20. The Spacing selected for both X and Y axes was 0.02. This tight spacing produced faunal density-distribution maps composed of 1,001 lines in both the X and Y dimensions. What these maps portray is an accurate spatial density per 0.25 m quadrant of different skeletal elements (basically, frequency of a given element per 0.25 m^2 unit space). As will be seen, the spatial geometry and configuration of faunal material from each secteur was radically different. The brighter the colors indicate higher element frequency in each map.

As figures 6.3, 6.4, and 6.5 reveal, there are two very different spatial patterns correlated with the differentially processed fauna in these two secteurs. Thompson (2011, 2014a) generated similar maps for each appendicular element, teeth, and axial elements but the results are redundant and very similar to the patterning seen in the figures above. The redundancy precludes inclusion of more maps here with essentially the same graphic patterns visible. It is obvious that bone was disposed of very differently in secteurs 190 and 202 II.22 at VBC, and that this phenomena has an evident spatial component. Two further figures, 6.6 and 6.7 provide further relevance to this patterning, one a photograph of material excavated in secteur 190, prior to piece-plotting and pulling the artifacts and the other, a ground-penetrating radar image of unexcavated archaeological deposits to the north of secteur 202. Figure 6.6 shows the roughly 2 m^2 circular feature visible in figures 6.3, 6.4, and 6.5, while figure 6.7 shows high-amplitude radar reflections of a semicircular feature that was partially excavated in field season 2009. Figure 6.6 clearly shows the circular feature visible in the bone density–distribution maps above. A section of an articulated thoracic vertebral column is clearly visible at center-right in Figure 6.6. Lithic knapping debris, assorted blades and some tools are also visible in the photograph at larger scales, interspersed with numerous small fire-cracked limestone and sandstone cobbles. While some artifacts were recovered in other areas of secteur 190 (most from a smaller circular features about seven meters south of the large circular feature) the deposit in Figure 6.6 produced a large majority of all artifact classes excavated in this 20 m^2 secteur.

Figure 6.7 is a ground-penetrating radar amplitude slice from an area just north of secteur 202, surveyed in 2006. This amplitude slice is 1 ns in thickness, showing high-amplitude radar reflections from a semicircular configuration

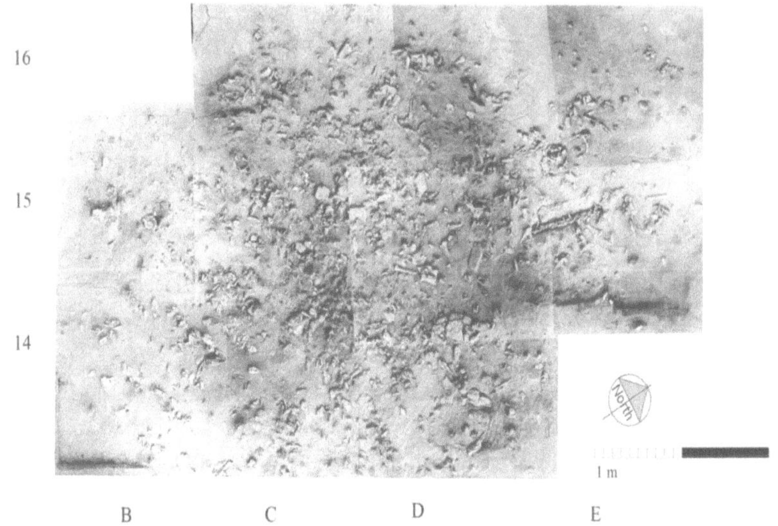

Figure 6.6. Field photograph of the newly excavated (in 1976) circular midden in secteur 190. Note the obvious circularity of the feature, and the jumbled, dense concentration of artifacts. Note also the articulated thoracic vertebrae at center-right of the photo. Photos of 1 × 1 m units around this feature were nearly devoid of any artifacts.

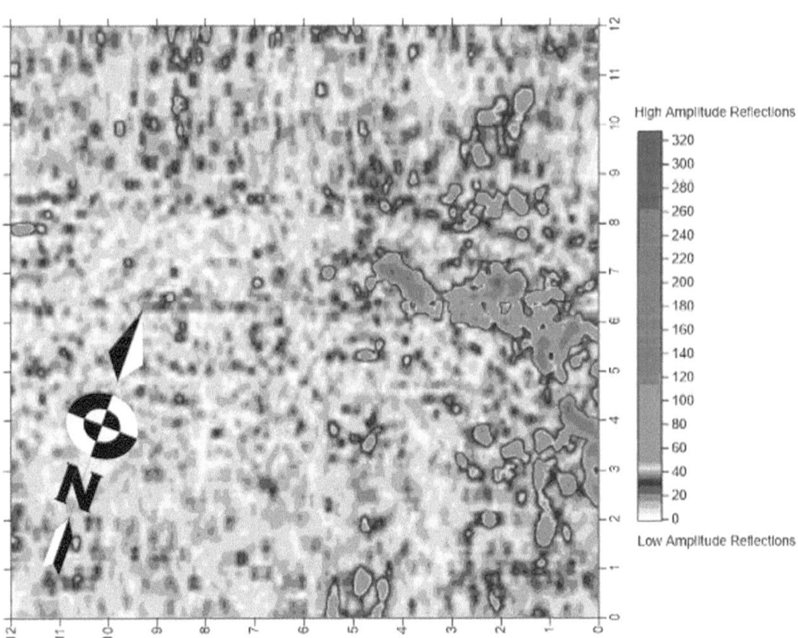

Figure 6.7. Ground-penetrating radar amplitude slice of area north of VBC secteur 202; using 900 MHz antenna, from 2006. Slice is 1 ns in thickness, at 24 ns time-depth. Note the semicircular feature clearly visible at center-right. This radar grid is 12 m^2, scale marks at 1 m intervals. A 2 m^2 section of the semicircle (the central portion visible at center-right above) was excavated and retired in 2009. Artifacts recovered included numerous bones, fire-cracked limestone and sandstone cobbles, lithic knapping debris, and several retouched tools.

of in situ archaeological deposits at 24 ns time-depth. In the field season 2009, a team from the University of Iowa excavated and retired a 2 m^2 portion of the semicircular feature, recovering numerous reindeer bones, an isolated but intact equid 2nd phalanx, fire-cracked limestone and sandstone cobbles, lithic knapping debris, several blade and bladelet cores, and some retouched tools. The 2009 field team used a total station theodolite to relocate the area in the 2006 radar grid above in Figure 6.7, and located the artifacts at expected depths just where the radar image suggested they should be.

It is unfortunate for the purposes of this volume that all of the VBC secteur 190 and secteur 202 material was recovered prior to the 2006 radar survey. Given the virtual absence of all artifact classes from secteur 190, except for those from the two circular features visible in Figures 7.4, 7.4, and 7.5, and the artifact frequency noted within those deposits (thousands), it is probable that a radar survey of secteur 190 would have produced high-amplitude reflections of those two circular deposits. The larger 2 m^2 circular feature in the northwest corner of secteur 190 produced, for example, a much higher density of artifacts per 1 × 1 m individual excavation unit than did either secteur 202 II.22 from or the 2 m^2 unit excavated from the area inside the radar grid shown in figure 6.7. The material excavated in 2009 from the area indicated by the 2006 radar survey (Figure 6.7) seemed broadly similar in artifact frequency per 1 × 1 m unit to that recovered from units on which the author personally worked at VBC in field seasons 1996 and 1997. In other words, in one's opinion a radar survey of potential unexcavated features at VBC similar to the circular deposit in the northwest corner of secteur 190 would be radar-visible (Thompson 2014). Likewise, unexcavated semicircular deposits at VBC, similar to materials in secteur 202, and to materials visible in the radar grid shown in Figure 6.7 would, in one's opinion, also be radar-visible. If that is correct, then perhaps at Verberie we have indices of differential faunal processing behaviors that could be visible in AGP surveys based on the scale, configuration, orientation, distribution, and frequency of archaeological material.

Would such putative and undiscovered, variably sized circular and semicircular archaeological deposits be visible at VBC? Figure 6.8 indicates that such features ought to be. In Figure 6.8, one can see a three-dimensional isosurface image of variably sized anthropogenic materials deriving from prior excavations. These were the features that were used above as examples of "false-positive" radar reflections. The visible features include a round concrete permanent site datum (for calibrating the total station theodolite), a backfilled trench approximate 1 m × 10 m, and a backfilled *décapage* approximately 5 m × 10 m. Other examples of radar-visible anthropogenic disturbances at VBC include plow scars and in some cases, individual root casts from sugar beets, shown in figure 6.9. The radar-visibility of subsurface material and even small-

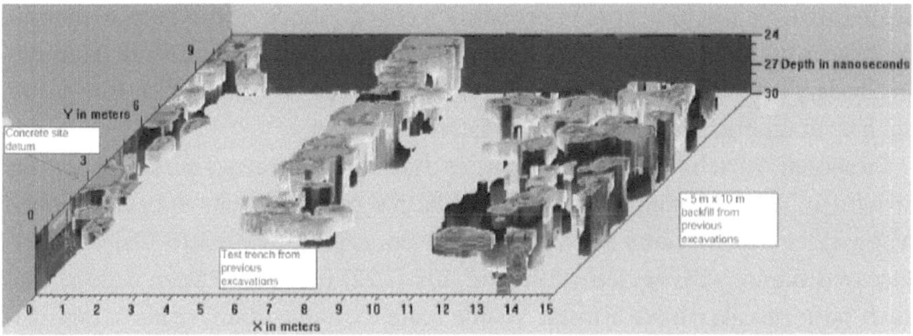

Figure 6.8. 3D isosurface image of radar-visible anthropogenic features at VBC. From left to right: Circular concrete site datum at left; 1 m × 10 m test or drainage trench at center from previous excavations; 5 m × 10 m décapage from previous excavations.

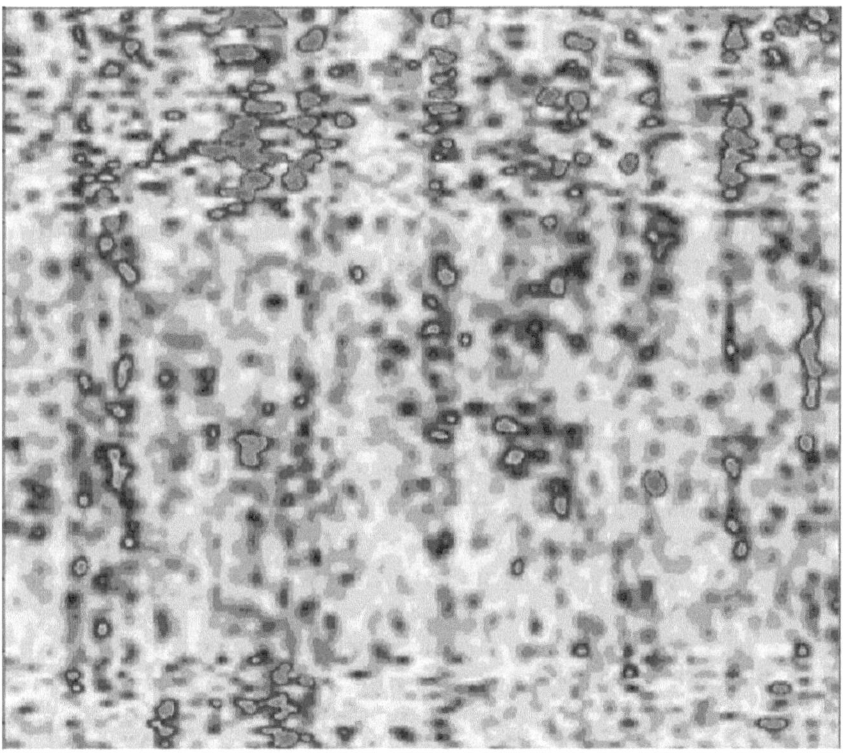

Figure 6.9. Elongated rows of plow scars at top-left, row of crop root casts top-center, individual rocks, and other small-scale soil disturbances and features at VBC. 900 MHz amplitude slice at 2 ns time-depth, 1 ns in thickness. Map is 12 m2.

scale intrusive disturbances is probably a consequence of the extremely fine, homogenous silt matrix at VBC. Since the 2006 survey used ground-penetrating radar, it would be extremely interesting to use another AGP technique, such as magnetic gradiometer or resistivity surveys.

Seasonal variability in prey species nutritional states, herd cohort age structures, and differential intensity of faunal processing have been invoked as possible explanations for the different bone-processing and disposal patterns visible in VBC secteurs 190 and 202 II.22. It has also been argued that such patterns may have a radar-visible scale, configuration, orientation, distribution, and frequency. There is, however, no guarantee that unexcavated components at VBC will conform to the patterning described herein. Let us assume that humans occupying VBC had perhaps used in some circumstances semicircular configurations of intact or mostly whole reindeer bones from geriatric carcasses with low economic utility. Let us assume further that the human occupants disposed of heavily fragmented, more youthful reindeer with higher economic utility in circular "trash pits" (for it seems that the large circular deposit in secteur 190 very closely resembles a trash dump). In either case, given the present AGP results from VBC, such patterns might be radar-visible to future AGP (or at least radar) surveys at the site. The precise structural characteristics of such putative assemblages could not be known until future excavations exposed the actual material indicated in putative future AGP surveys. The larger point is that the *physical presence* of unexcavated archaeological deposits ought now to be demonstrable at VBC based solely on AGP survey, and that even prior to any excavation something might be knowable about them in human behavioral terms, however vague.

A further curious fact about VBC secteurs 190 and 202 regards the following: there were three hearths located inside the 20 m^2 arbitrary boundaries of secteur 202 while there were none located inside secteur 190. Therefore, coincident with the patterning already described, we have another index—presence or absence of hearths—to add to our consideration. It appears that archaeological deposits in the proximity of hearths at VBC may portray semicircular configurations while at least one located at a distance from any hearths evinces a tight, circular arrangement. One has visited this site on five separate occasions (1996, 1997, 2006, and twice in 2009) and each time one has observed outstanding faunal and lithic artifacts lying on the surface, normally after plowing and planting or harvesting of the agricultural field located above it. In one's opinion, there is a strong likelihood that much or most of this material is dragged generally east to west from plow activities and planting furrows. The eastern portion of the site may be the least-sampled. The Pleistocene point bar on which VBC is located is about 25,000 m^2 in area while approximately 600 m^2 have been excavated to date, representing about 2.5 percent of the total area available for human occupation and exploitation (Enloe et al. 2007; Thompson 2011, 2014a). Thus, the totality of inferences drawn from Verberie rest upon a miniscule sample of the total area the site may encompass. How likely is it that the site boundaries have been established? One does not intend to sing a paean to future research at VBC, which is nonetheless clearly warranted, so much as it is to state plainly that for all that is known about a particularly valuable site, this uncertain knowledge is based upon rather a small foundation and progressively more data is being erased annually as the field is cultivated. How much of an incompletely excavated site must be excavated to characterize the entire site? How much of a valuable site can we afford to lose before it is completely destroyed inadvertently?

A comprehensive total site survey at VBC using AGP would be useful to discover whether we can "see" further evidence of the two material patterns noted above repeated over unexcavated space at the site, or whether other material patterns can be found. Since the two spatio-material patterns (round disposal pit perhaps indicative of spring occupation, and semicircular pattern possibly indicative of fall occupation) located to date may have seasonal ramifications, the location of further patterning might change even more our traditional interpretation of the site as a one-season fall logistical butchering camp. Use of VBC over multiple seasons of an annual round could perhaps indicate a more mobile forager type of Magdalenian adaptation, with the site used more dynamically over a wider seasonal range of resource variability (Thompson 2011; 2014a) This has implications for how we interpret Magdalenian mobility and settlement patterns. It may be that occupations over multiple seasons at VBC indicate that the Magdalenians themselves were more "mapped on" to the Paris Basin than previously thought.

NOTES

1. (http://mysite.du.edu/~lconyer/misc_software.htm).
2. (http://mysite.du.edu/~lconyer/misc_software.htm).

CHAPTER 7

Failure at 13DB674 in Iowa or Redefinition of Success?

UNFORTUNATELY, WE DON'T REALLY HAVE AN ACTUAL DEFINITION FOR either "success" or "failure" in terms of AGP applications in archaeology, so we have to use subjective ones, such as, "It worked and produced useable images," or "It appears partially to have worked," or other verbal formulations that involve imprecision and appeal to the relative quality of imagery obtained from AGP surveys. There is a substantial learning curve involved in utilizing AGP, and one's experience is probably the single biggest factor in measuring successes versus failures (Conyers 2004). In this chapter, one will describe and highlight what one interprets would be regarded by most AAs and GAs as a relatively "unsuccessful" application of AGP technology at a North American Archaic site in Dubuque County, Iowa, USA (site number 13DB674, hereinafter abbreviated as 13DB). I will present some descriptions of what the AGP survey "located" and what was exposed in excavation similar to chapter 7 above, and end the chapter with a summary of the possible causes of our perceived "failure" at the site.

The site 13DB674 was subjected to a salvage excavation in fall 2012 by Tallgrass Historians, a small, Iowa-based CRM firm. The dimensions of this site were quite large (over 200 m^2) and were spread over multiple loci. After one's involvement with the project had ended, one was notified by the firm's co-owner that a diagnostic North American Archaic projectile point had been excavated, indicating that occupations could be expected to date to that period. One has previously described this application as an "unsuccessful" application. What one means by that is not that good imagery was absent, but that the radar images were not characteristic of the *archaeological* material. This project acquired abundant quality subsurface data at 13DB674 that were indicative of cryogenic and fluvial material patterning as opposed to behavioral. The radar survey produced images of the channel bottoms of small rills and gullies in plan view (figure 7.2), and some excellent radar profiles (figure 7.1) of these small channels in cross-section. These and similar images from the Dubuque site will constitute an important comparative dataset, and potential

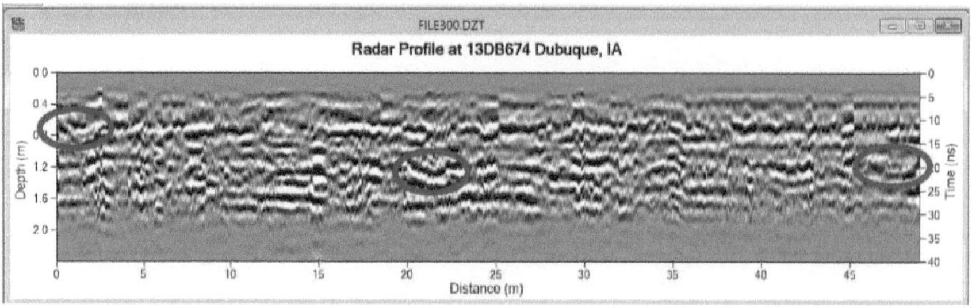

Figure 7.1. 400 MHz radar profile at 13DB674 Archaic site in Dubuque, IA. Note the nearly diagnostic distorted u-shaped fluvial channels inside ovals. There are other smaller rills and small gullies also visible as u-shaped reflections visible in the profile that have not been ovaled. These features were virtually invisible to the naked eye and photography.

cautionary examples, because they appear superficially to resemble potential buried archaeological deposits. A particularly large feature 13DB674 that produced excellent high-amplitude reflections was found to be nothing more than a wide slackwater deposit located along the margin of the small channel near the site, known as Catfish Creek. In amplitude maps and time-slices this feature looked like images of intrusive disruptions (from trenching, excavating, and backfilling) and in situ archaeological deposits gathered at Verberie. Instead, the composition of this feature was of fine cross-bedded silts and fine sand, and was thus completely a sedimentary feature. No artifacts or even many rocks were found in excavations of this feature.

In Figure 7.2, we can see the sort of radar reflection patterning that was found to be typical at 13DB674 (high-amplitude reflections of buried material showing various linear and sometimes circular configurations). There were *no* individual rocks found of sufficient sizes to produce these reflections during excavations, no anthropogenic structures or hearths, with highly dispersed and discontinuous artifacts indicative of a heavily eroded and redeposited site. The high-amplitude reflections can be seen to portray linear patterning (generally at diagonals upper left to lower right and vice versa), and, coupled with the u-shaped reflections visible in profile, show the physical remains of some of the erosional culprits. In fact, at 13DB674, the densest radar reflections appear to be of preserved erosional surfaces (Artz 2014; Thompson 2014b).

The AGP survey at 13DB674 indicates just how valuable unpublished AGP applications can sometimes be, especially when the features that produced radar reflections above (rills and gullies located across an eroding floodplain at multiple depths) were nearly invisible to the naked eye and superficially resembled radar images from sites that feature actual buried archaeological materials. Although the imagery from 13DB674 apparently only captured natural patterning, this provides a handy future reference for our comparative

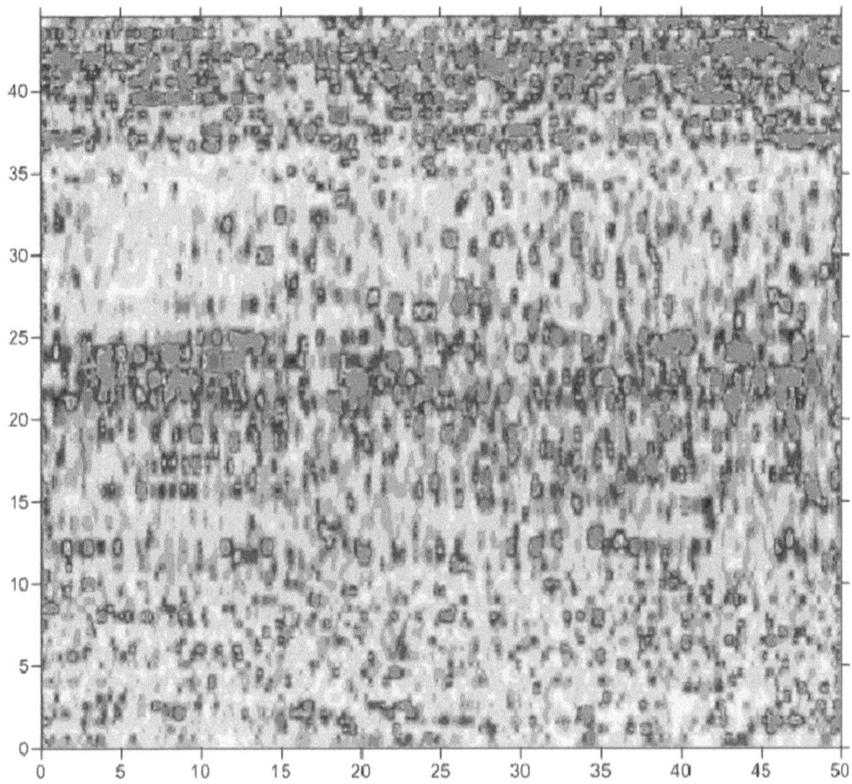

Figure 7.2. 400 MHz radar amplitude-slice at 35 ns time-depth 1 ns in thickness from 13DB674, Dubuque, Iowa. Note high-amplitude reflections of approximately meter-sized configurations of what were ground verified as vague sedimentary structures (the bottoms of infilled fluvial channels).

collection. It appears that human occupations of this site during the Archaic were either extremely infrequent or temporary, or as seems more likely given the number of fluvial features we found, the archaeological material has been pervasively eroded and redeposited. Had this data not been acquired we might be ignorant of how wide slackwater deposits or rills and gullies look in the setting of the Midwestern Driftless region, and that is valuable in its own right. Whereas the radar antenna used at 13DB674 was 400 MHz, I don't think the use of a higher frequency antenna would have produced better images. They may have been of higher resolution, but the images we got were excellent. It's just that they didn't show archaeological materials.

The AGP surveys conducted at 13DB674 present us with some fascinating examples of site-formation and post-depositional taphonomic processes at least partially visible in radar profiles and amplitude-slice maps. In other words, the radar profiles and maps from 13DB674 revealed some subsurface characteristics that might suggest or from which we might infer an absence of spatial integrity even if anthropogenic materials are located. For example, based on the AGP surveys at 13DB674 in 2012 one suspects that for any project area the presence of small u-shaped fluvial channels spotted in profile especially, but also even in plan view, cutting across a site and situated at many different depths, should be a potential red flag. In the case of 13DB674, approximately 150 m^2 of the site was systematically surveyed over three separate grids located in areas that were not augur-tested during the much earlier phase II activities by which the site was originally located. Much time, effort, and money was expended using the radar results as means to emplace full excavation loci, which proved to be essentially fruitless. The majority of diagnostic artifacts were not even excavated inside the radar grids, but were unearthed instead within spatial units (1 × 2s) that were chosen at random.

So how should we summarize the redefinition of success at 13DB674? Conyers (2012, 2013) suggests that a constant field and even lab practice should be to analyze AGP data (at least ground-penetrating radar data) by monitoring differences and similarities seen in profiles and in plan-view. Undue focus on either can easily botch a field application. At 13DB674, the radar profiles clearly revealed the presence of pervasive channelization that cut across the body of the site, however, the mistake of focusing undue attention upon amplitude-slice plan-view maps of the radar grids, trying to find "hot spots," resulted in overestimating the site's spatial integrity. Finally, this AGP application at 13DB674 also suggests that, in the haste to "get things done," we frequently mistake rapidity in conduction fieldwork for "efficiency." Had the crew at 13DB674 (and this includes the crew supervisor, consultants, and field staff) taken opportunities to stop, compare preliminary results of the radar surveys, soil corings, and excavations, and communicated, we might have learned much more about this "site" and actually saved money. Whereas the field mentality was "push!, push! dig! screen! ASAP!", a slightly less time- and money-centric and totalized professional strategy could have saved both time and money. Artz (2014) and Thompson (2014b) both conclude that 13DB674 is not so much a site as it is the eroded and redeposited remnants of many sites from upslope. In 20/20 hindsight, which is always clearer than dynamic real life, the data argued even from the inception of the 13DB674 project that the "site" was in reality anything but.

CHAPTER 8

Framing Testable Humanistic Hypotheses for AGP

THIS IS PERHAPS THE MOST DELICATE AND DIFFICULT PART OF THIS BOOK: how do we frame testable hypotheses for AGP research that feature humanistic content or merit? Most traditional AGP research has not been concerned with framing humanistic hypotheses. Instead, the goals appear to have been framing a basic hypothesis of "does it or will it work at Site X"? A distinct avoidance of humanistic themes is apparent from even a brief perusal of the primary AGP literature. On one hand, this results from an overt focus on technical considerations and the "arms race" to publish various "successful" applications of various methods at sites. Yet, it also derives from a lack of focus on human themes, as though anthropology and social science are subordinate to, or at least less important than, "hard science," such as geology and geophysics. However, even the purest geophysics have rarely been devoted purely to "science," being devoted instead to prospecting for hydrocarbon or other mineral deposits, which as applied to for-profit oil prospection has utterly nothing to do with scientific goals. In application to archaeology, AGP, willingly or not, enters into the purview of the social sciences and humanism. It comes with this territory.

If we regard science as merely the process of testing falsifiable hypotheses, then even in that narrow moment we can make some progress. What we need to do is to frame such hypotheses correctly. I think this is really rather simple: we can frame hypotheses in terms of, "Is there AGP-visible evidence of *x* at a given site?" What this forces us to do is to know beforehand the kinds of things we might expect to find or not to find at any given site. In other words, we have to reference our comparative collection of AGP images to ascertain what we would expect things to look like if we found them, across the widest range of variability possible. If we don't, we're doing no better than random shovel-testing and may as well forgo the expense of AGP techniques.

One has two main rules for AGP: Rule 1) AGP *always* works and does what it is supposed to do unless one has defective or malfunctioning equipment or users. *Results* can be unusable due to interference or soil/sediment conductivity, combinations of such factors, or inadequate data processing, but that is not "failure."

Rule 2) Don't use AGP unless or until one is willing to invest requisite time, effort, and money to process raw sampling data into usable media and situate the AGP images into a larger coherent research design (in tandem with technical analyses of artifacts, spatial archaeology, etc.). Stand-alone AGP research is great, but we can more fruitfully apply what we learn with such techniques by applying them in tandem with other archaeological techniques.

A SET OF BASIC STEPS FOR AGP RESEARCH FRAMING

Here is a basic outline of steps and questions to ask and to answer for framing AGP research in humanistic anthropological terms. First, have we located a site via pedestrian survey, AGP survey, or direct subsurface testing? What do we wish to learn about the site purely in terms of AGP techniques? What kind of site does it appear to be in scalar terms (is it a monumental or small-scale site)? Do we know the site's areal extent? What kind of site can we expect it to be given survey data and other indices? What sorts of archaeological material features appear to be present? For example, an awareness of the kinds of buried features expected can help determine AGP techniques to employ, along with survey transect spacing. What aspects of being or behavior might be AGP-visible at the site? Does there appear to be variability or homogeneity in terms of material scale, orientation, configuration, distribution, or density? Can we observe via AGP variability across horizontal or vertical space at the site?

Many AGP professionals (too many to cite appropriately) wisely suggest that we should use multiple complementary AGP techniques at sites. Yet, not only complementary AGP methods should be used alongside one another, other forms of archaeological technical analysis should also be advocated. Obviously this assumes that ground verification through core, augur, or excavation testing have to be used to substantiate inferences drawn from AGP data. How did archaeology learn how to accommodate technical analyses of fauna, lithics, and site integrity? We should embed AGP in tandem with other analytical methods for extracting understanding from the archaeological record. This must be done to "test" the impressions we gain from AGP applications. Using AGP data alongside other powerful methods, such as faunal or lithic analysis, provides other means for testing the data.

Testable humanistic hypotheses can be framed in terms of falisifiability, such as "one hypothesizes that Method X will 'work' at Site Y." Or such hypotheses can be framed using multiple AGP techniques. For example, we could frame such hypotheses as follows. "There are AGP-visible aspects of behavior at Site

X," is a perfectly testable hypothesis. Or, "there is evidence of change in material scale, configuration, orientation, distribution, and density across horizontal or vertical space at Site X" is also a valid and testable hypothesis. Of course, such hypotheses assume that basic AGP feasibility-testing for utility has been performed alongside coring or auger-testing, along with pedestrian survey, to learn about substratal material composition and properties. Don't head out to the field with AGP equipment unless it has a probability of working. A proper fit between site dynamics and AGP techniques has to be estimated. Hypotheses framed as above can also lead to humanistic anthropological inferences when combined with the totality of evidence available from excavation and technical analyses in instances where excavations occur. If complementary AGP methods can be used productively, then combining AGP in tandem with established anthropological archaeology techniques should reveal much more. I do think geophysical anthropologists have to adapt to norms of humanistic anthropology, and I also think that these are some ways we can start to implement the adaptation.

DISCUSSION

At this point, we have plenty to discuss. Quite apart from the varying physical, structural, chemical, and other dynamics that affect AGP applications at individual sites, we have some less objective but equally, if not more, crucial matters at issue. First, how should we define relative AGP "success" vs. "failure"? Since we each bring different expectations to the table, one suspects we will have many different definitions of both AGP success and failure. So, we need to talk definitions a bit more, one thinks. Does AGP "success" – pretty pictures of the subsurface in whatever or however many dimensions, profile, or plan? Is "failure" a lack of such pretty pictures? Judging from the current AGP literature, it appears pretty pictures still have a lot of influence on publication norms. Is "failure" the production of ambiguous results, whereby "success" is achieved in telling diggers exactly where to dig and what to ignore? Assuming a more or less constant increase in AGP technical and methodological sophistication, would success occur by literally "seeing" underground? Or could it be that our conceptions of both "success" and "failure" need to evolve in order to get the most anthropological currency out of the research? One leans progressively more toward the latter, because one thinks our constructions of success and failure in relation to AGP remote sensing are quite limited. Such definitions should change with time as we learn more about anthropological AGP.

In order to be consistent with the anthropological perspective from which one approached this volume, one would first ask what makes either a successful or unsuccessful anthropological study, perhaps premised upon something tangible like culture change? In order to frame such a study, an anthropologist might set about developing expertise with a particular culture, existing in a particular geographic culture area, through use of the guiding paradigm or anthropological theory of cultural relativism. Such an anthropologist would probably then construct a research design premised upon the established anthropological method of participant-observation enacted over multiple instances over some extended period(s) of time. Testable hypotheses would be formulated whereby culture change would be refuted or verified. Data would be gathered, results quantified and analyzed, hypotheses confirmed or refuted, and conclusions rendered. Can such methodology inform us for AGP research? Obviously, we cannot study culture, or even *a* culture, at least not directly, through AGP.

One submits, however, that given proper framing and within certain limitations we can research aspects of culture through such techniques. For example, one seemingly valid end-stage anthropological generalization from AGP data at the site of 13DB674 would be that for a variety of reasons this particular site was not occupied permanently or frequently during the North American Archaic period. In particular, the Archaic inhabitants of 13DB674 appear not to have engaged in much or any on-site processing of faunal material, especially in relation to the number of lithics recovered at the site. Yet, we had to pass through several layers of data acquisition, excavation and ground-verification, lithic and (exceedingly minimal) faunal analysis, before such a conclusion was possible. Likewise, at Verberie, although a smaller total area was surveyed, and the results are less salient therefore for the total site area, it was also only after an entire litany of studies (pedestrian surveys, decades of excavation and technical analyses, along with AGP survey) that one could suggest that some aspects of behavior (in this case, differential seasonality) are visible at that site via AGP. Is this really anthropological? It certainly isn't sociocultural or applied anthropology, but yes, it is anthropological, because the issue is human behavior and its visibility or invisibility via certain AGP methods at some sites. So it's definitively anthropological.

One extremely valid reason for the preponderance of "pretty pictures" in successful AGP application is sheer usefulness, as from some of the imagery obtained at VBC of both in situ unexcavated material and disturbed sediment from prior excavations. However, such graphics are also sometimes ambiguous or misleading, as we also see at VBC, and at 13DB674 as well. In fact, with regard to 13DB674, none of the "hot spots" that seemed to be indicated from AGP produced much in the way of a visual guide to where to dig or what to avoid digging. None of the diagnostic artifacts from the Dubuque site were excavated from the radar grids. Given the very small scale and dispersed distribution of artifacts recovered at 13DCB674, it is unlikely that any artifacts appeared in the GPR data from the site. This is in contrast to the recognizable middens or scatters that are clearly visible and identifiable at Verberie in GPR data. So imaging success especially in portraying buried larger-scale materials is with us to stay, because it's powerful. But are images like those from 13DB674 powerless or completely "unsuccessful"? One doesn't think so. They accurately portray something belowground, but they just require a bit more effort to pull us up from the mud, and to see what they have to reveal for us. One is still learning how to do anthropological AGP. We all are. Anthropological archaeologists also need to buy and use AGP technology more.

Is the fact that AGP involves subjectivity inherently *bad*? One recalls stern conversations with a CRM professional who was visibly disappointed with an AGP survey and criticized it as "apparently fraught with subjectivity." What isn't? Does subjectivity = unsuccessful, given AGP's currency as an archaeological technique? Is there any archaeological technique that is purely objective? In other words we have here a basic question about science and the nature of reality, and we can ask it of many things. Is faunal analysis or any technical analysis inherently objective? How many faunal analysts analyze their datasets the same way? Is reproducibility really germane to the discussion at the level of any technical analysis? Interpretations of any faunal dataset will vary, for any number of reasons. We also have to remember that archaeology is not, strictly, "science," whereas geophysics is. Expertise does matter, and this has to be accounted for in any responsible research framing. One thinks we can get at some basic anthropological relevance in pretty short order if more of us participate.

At a recent AGP symposium held at the 2010 SAA meeting in St. Louis, Patch and Gregory (2010:1) noted some useful concepts at the level of the site: "What components are represented? What is the basic function (site type)? What is the internal structure of the site? Are different feature types present? What is the spatial (both horizontal and vertical) distribution of features? Are probable burials present, and, if so, what is their spatial arrangement? Are different components present? Can these be identified and interpreted from the radar data? What do these features tell us about human behavior?" Those are great questions for applying anthropological AGP in the CRM realm. We should also ask, for any AGP survey, how does human behavior condition material visible to AGP techniques in manners that allow us to draw some inferences? Human behavior impacts materials in a process, producing aspects of being that may be visualized at later dates.

What again are aspects of being? By this one simply means human conditioning effects on material distribution, configuration, scale, and orientation that can lead to inferences about human behavior. What can we learn from a site? Via AGP, we can learn how differentially to "weight" tests of multiple areas quite quickly at sites. Not all artifact-bearing areas at a site are equal. Differential time/money constraints might make working less dense areas more effective in the short term before returning for subsequent comprehensive tests. That's sort of an inversion of the common orthodoxy which might have it that we use AGP to find "where to dig," by a sort of artifact or reflection density-analysis. We'll choose to dig where it seems from subsurface reflections as though we have a higher relative probability of digging up some good stuff. Yet, maybe we don't wish to look at a valuable archaeological site in such a way. Perhaps we might like to plan excavations at a noteworthy site differently and allow for how tests inform us diachronically and alter our research designs. AGP offers us the ad hoc ability to introduce finer resolution of estimates for excavating sites, as well as to emplace these within larger archaeological landscapes. Thompson et al. (2011) additionally suggest four categories of human spatial use that we can address geophysically: Category 1: Identifying construction variation in terms of the built environment; Category 2: Identifying continuity/discontinuity in the use of space; Category 3: Identifying natural and cultural modifications; Category 4: Identifying regularities in the use of space and architecture at the regional level.

In effect, AGP gives us better mechanisms than pedestrian survey for formulating viable research designs. This is especially true for those instances in which we might need to learn more about a site before spending time and money on other available testing techniques. This does, of course, offer methods for incorporating cooperative, multidisciplinary participation by workers from differing specialized disciplines. One thinks, however, that we should be focused not primarily upon various technical parameters (equipment settings, wave propagation velocity, etc., which while important don't actually conform to anthropogenic causality but rather to structural, chemical, and physical attributes) but instead upon the interpretation and translation of subsurface data into widely useable archaeological and anthropological information. Lest my colleagues flay me for being naïve, one is perfectly aware that material attributes condition our view of the subsurface and also of everything in it. One submits, however, that much of the AGP literature focuses too much on the one and not nearly enough upon the other.

We can, however, collaboratively and mutually participate in forging new AGP methodology, new publications, and new research foci to address shortcomings. We can all derive benefit from anthropological AGP. All it requires of us is a willingness to develop new methods for AGP and to teach ourselves new ways to think about AGP equipment, other technical analyses, and archaeology in general. Not all sites will be like Verberie, at which a highly fortuitous collocation of human behavior, animal bones, fine, homogenous silty sand, and gentle flooding resulted in outstanding spatial integrity of archaeological deposits. That those same archaeological deposits appear, hypothetically, to show AGP-visible indices of human behavior is perhaps quite rare in the archaeological record. We don't know. We're in no position to say. The structural realities in the discipline of geophysical archaeology and its disciplinary literature have somewhat constrained the sort of anthropological premises we might wish to find. We're now ready for a much deeper focus on people and human behavioral signatures perhaps visible in the archaeological record using AGP. One thinks what we need is just different framing.

McGuire (2008:7) has established much of a material means by which to interpret archaeological phenomena: the four C's; Coherence, Correspondence, Context, and Consequences. In McGuire's construct (2008:7), Coherence relates to the relative harmony between theoretical and logical positions; Correspondence relates to how our interpretations fit with our observations of the real world; Context describes our sociopolitical and cultural matrices; and Consequences, how our interpretations of the archaeological record function in the real world. For if AGP is to be mainly concerned (as it must be) with material means of assessing the archaeological record, then we should employ a wide range of thinking about material. If anthropological archaeologists intend to use AGP productively, then they have to bridge the gaps too, to find avenues of better communication and understanding. We have to avoid dead ends. One way we can avoid dead ends would be to establish some very basic goals and parameters by which we incorporate AGP into anthropology, to which the final chapter is devoted. One won't present "the way" to do AGP anthropology-style; one is instead presenting some ideas for how to do this in the hopes that others will also subsequently offer their inputs, so as to make this a legitimately multidisciplinary avenue for the production of informative and useful research.

One of the ways that one approaches anthropological research framing for archaeological geophysics is to use analogues. AGP is nothing more than another tool we have at our disposal to investigate aspects of the past. AGP can be incorporated into our toolbox nearly like faunal analysis, or lithic analysis, provided that we establish the theoretical parameters whereby we do so. How do faunal and lithic analysts use comparative collections in their research? What conclusions do they reach by using such references? How do they reach their conclusions? What methodological and theoretical bridges are built to allow conclusions? These are just a few of the questions to answer. What I will now describe is how one used AGP as a middle-range research tool, including some personal examples and ideas. This is not meant to be authoritative, just instructive and helpful. Above all, one hopes to engage others in a constructive dialogue so that we can improve our methods of use for this powerful battery of technologies.

In 2005, during one's dissertation research, a collaborative team of French and American archaeologists developed a program for AGP application at the site of Verberie. In this case, the particulars suggested that it would be a good site at which to use ground-penetrating radar, due to the low clay/high sand content of the site matrix and the knowledge that at least some unexcavated material exists at the site. At that point in time, few applications of any AGP techniques had been undertaken at Paleolithic sites. One of the things we wanted to do was to assemble our own database of AGP imagery specific to the Upper Paleolithic at Verberie, to demonstrate how things look underground, to build our own "comparative collection" of images of unexcavated, subsurface things that we could use as a reference. One compiled images of as many imaged subsurface things as possible, putting them into various folders and directories on PCs, flash drives, and CD-ROMs so as to have a comparative database, in case we needed to reference back and compare newly acquired data to earlier data.

As a faunal analyst, the best thing one can do for my colleagues is to assemble the biggest comparative collections that one is able. In order to analyze the widest variety of faunal assemblages one needs to have at their disposal the widest variety of skeleta that one can assemble. No faunal analyst can identify every bone found at a site, especially when the bones are fragmented. Even in those instances in which faunal assemblages are relatively complete, a comparative collection is necessary because animal bones vary according to many individual variables. There are sex and age differences, size differences, health status differences, faunal treatment differences premised upon elements of human agency and human preference. There are any numbers of variables that condition animal bones at archaeological sites. In order to make sense of new assemblages one's duty is to build the largest comparative collection possible. The same holds true for AGP images, because without a ready and accessible set of images for what things look like underground, what sense can we really make of any images we get from any particular AGP survey, barring complete excavation of a site? If we have nothing with which to compare new data, then what sense can we actually make of it? What do new images "mean" if there is nothing we can reference that looks similar or different?

This is not a new phenomenon, in that lithic replicators have the same problem. In order to make sense of the debitage we produce during lithic replication studies we must have a ready sample of actual stone tools on which to base our conclusions from our own, original knapping efforts. It's not good enough that the things we might make when knapping look like actual lithic artifacts. That's the easy part: what is difficult is comparing them to actual examples of prehistoric manufacture and extrapolating from those some aspects of being, some real human behavioral attributes that make sense. With respect to AGP, and especially humanistic anthropological applications of it, we need to compile as comprehensive a set as possible of both real and simulated comparative images of subsurface archaeological features. We have already a very substantial set of comparative images from actual monumental and other large-scale archaeological phenomena, such as castles, bridges, foundations, and churches. We lack, however, a similar set of simulated phenomena, precisely because it would be impossible to replicate such megastructural items. For small-scale things, such as hearths, middens, and the like, we need a much more complete and open-source, sharable set of comparative images, which opens up many possibilities for experimental research and simulation/replication studies. In fact, lacking a comprehensive database of these types of comparative images, it will be difficult to get the most out of AGP. One notes that the performance of such research would be a great way to involve both undergraduate and graduate students in an important aspect of AGP: middle-range research.

Conyers (2012) suggests that anthropological archaeologists and geophysical archaeologists work toward more collaborative endeavors, and details some ways we can approach them, indicating that one way we can do this would be to get more geophysicists to think and to train anthropologically. Learning anthropologically would include learning new reference frames for thinking about people. Such a process should also invite more anthropological archaeologists to reduce their ignorance about the possibilities AGP can offer, but I think this cross-fertilization or hybridization process must by necessity place the onus to adapt squarely upon both geophysicists *and* anthropological archaeologists. The program should be to meld aspects of both geophysics and anthropological archaeology into a coherent body of thought, including sets of methods and comparative subsurface image collections to use in application to anthropological analyses of archaeological sites. Geophysical goals can be met, but geophysics is not really why people analyze archaeological sites, it is *how* we can learn to analyze sites. We conduct archaeology in order to learn something about past peoples, which we can do by finding new ways of using AGP technology.

Conyers (2010) suggested that it is now time to stop viewing geophysical archaeologists as mere appurtenances of archaeology since AGP techniques and methods have evolved to such a state that they can offer substantive contributions to archaeological method and theory. This is true. AGP needs to adapt to archaeology and anthropology as much as anthropology and archaeology to geophysics. The main interest in archaeological sites should be to address anthropological theories, including those of salience to geophysical archaeologists, to answer humanistic and scientific questions. Another goal should be to get archaeologists to think about new ways to incorporate

AGP into their basic research designs. This process should therefore involve geophysicists learning anthropology as much as it would mean anthropological archaeologists learning new ways of thinking about geophysics in application to their own research.

Anthropological archaeologists need to know how archaeological deposits "look" underground so we can identify more of them in unexcavated areas or rule their presences out. In the case of Verberie, we assembled a large comparative collection of radar images of replicated small-scale open-air phenomena because we weren't certain what the field applications in the Upper Paleolithic of France would elicit. We were trying to use radar to find unexcavated areas but how would we recognize them if we saw them? We also need to rule out false-positive reflections of previously disturbed substrate from decades of prior excavations. In essence, this led to the creation of new methodology both to identify actual archaeological deposits and rule out false-positive sediment disturbances that mimicked them.

AGP should continue to demonstrate its relevance to larger issues of anthropological theory, social theory, culture, context, evolution, and all the other themes that represent the purview of archaeology as it is practiced today. The power of AGP technology is, however, most relevant to anthropology insofar as it helps to answer questions or to solve problems of humanistic or anthropological merit. A consistent and productive program of interdisciplinary research has to address issues of disciplinary merit for both geophysics and anthropological archaeology. Is there an avenue for publication of genuinely interdisciplinary research in AGP? The publication norm typical may be changing, as a wider range of scholars publish AGP research in a wider range of literature. The bias toward images of huge materials at huge sites is still surging forward, however, and shows little sign of stopping, although new volumes of *Archaeological Prospection* have begun to include more experimental applications, which is an encouraging find. There is, however, still an absence of the kinds of experimental, simulation studies that typified middle-range zooarchaeological research, lithic replications, and material refitting studies, where novel ways of thinking, methods, and experiments were developed to meet the needs of archaeologists studying humanistic phenomena. There is a vast and informative area of experimental AGP research to perform.

CHAPTER 9

Ideas for How to Do This

As someone who is primarily a zooarchaeologist, one would advocate some of the following methods for making AGP more relevant to anthropological archaeology. Here's what one would like to see it do. First would be a dedicated program of research devoted to imaging simulated small-scale archaeological features such as hearths and middens. Most of us have access to yards. It should not be difficult to use our own yards in some research programs to excavate controlled openings and to make simulated archaeological features indicative of the sorts of small-scale features one described above, in many different matrices and at variable depths. For example, one would first acquire some peripheral storage media, such as flash drives or hard drives, and set them up with basic folder menus. Such menus could include basic folders such as "real" vs "simulations," and "raw" and "post-processed" data. Data collected from real or simulated sites and features could be easily placed into folders such as "hearths," "middens," "domestic structures" or other such folders as needed. The goal would be to compile the biggest and most representative set of AGP images available (ideally published open-source for general use by nonproprietary personnel) for comparison/contrast with images obviously derived from actual archaeological sites.

It is difficult, in lieu of substantial experimental imaging of simulated material configurations, to impose an arbitrary range of patterning that ought to be

Table 9.1. Material Patterns Potentially Visible to AGP

Small-Scale Patterns	*Large-Scale Patterns*
Regular Shapes*	Regular Shapes*
Irregular Shapes**	Irregular Shapes**
Surficial Expression	Surficial Expression
No Surficial Expression	No Surficial Expression

* lines, circles, polyhedrons, regular dimensions
** irregular shapes, irregular dimensions

visible to AGP technologies. We actually have to simulate a very wide range of variation in order to make this work.

In terms of developing promising research designs, one would propose the following possible topics of inquiry, involving both experimental, simulated projects as well as actual archaeological applications. In terms of simulations, virtually a limitless array of possible research awaits the courageous scholar; one writes "courageous," because one cannot predict how the discipline will receive such research in peer review. There could be some resistance, but this is certainly a fertile area. One separates such research into categories based on material scale, including simulations of hearths, middens, pits, burials, wells, privies, and any number of other small-scale features and structures. Example of larger-scale simulations could include footings, walls, foundations, ditches, other kinds of structural ruins (collapsed roofs, collapsed walls), and streets. For examples of productive real small-scale research agendas, one would suggest a focus in the United States on AGP applications at more Paleoindian, Archaic, and Woodland sites, with special emphasis on imaging of kilns, hearths, and other small-scale features. Large-scale research of actual features and structures could obviously include mounds of all sizes and types, as well as Adena, Hopewell, and Mississippian sites. Table 9.2 below summarizes this information.

As noted above, engaging in research of the types described, especially the experimental simulations, will require some ingenuity and independence on the part of the researcher. Perhaps the more influential and prestigious of the archaeological scholarly journals will come around with time and become more receptive to publishing such material, whereas the fact still remains that the utility of such research and its usefulness to the discipline need to be demonstrated in further books of this kind, but also in grant proposals, technical articles, open-source e-publications, and other media. Furthermore, for the

Table 9.2. Potential Avenues of Productive AGP Research

Simulations	*Small Scale*	*Large Scale*
	hearths	footings
	middens	walls
	pits	foundations
	burials	ditches
	wells	ruins
	privies	streets
Real	*Small Scale*	*Large Scale*
	Paleoindian	Mounds
	Archaic	Adena
	Woodland	Hopewell
	Kilns	Mississippian

suggestions detailed herein to take maximum effect, both AA and GA need to make some changes relative to one another, as one suggested above, in order to facilitate increased and improved dialogue but also to reduce the potential for miscommunication.

Some examples of productive future applications for AGP would include analysis of surficially "invisible" places (Fontijn 2007) and structured aspects of being, including community, political identity, and social memory expressed materially and spatially at Moundville, Alabama (Wilson 2010). For Bronze Age occupations of the Neteherlands, Fontijn (2007) has described a class of sites known as (surficially) "invisible places," or unmarked deposition sites for metallic valuables, with clear ideological ramifications in the differential spatial and landscape deposition of metal items (swords and other martial items deposited in or near rivers and stream confluences, nonmartial metallic valuables deposited in peat bogs). AGP surveys of such known locations, and landscape surveys between them, could prove to be particularly informative of further potential surficially invisible places. For the Moundville, Alabama, site, Wilson (2010) has proposed that sociocultural phenomena of community, group identities, and social memory is expressed in terms of the physical and structural configuration of mounds and intervening areas. If Wilson (2010) is correct, then comprehensive AGP survey of that and further mound sites could help inform us about surficially invisible aspects of cultural phenomena that were structured and expressed spatially. We seriously need also to try and calibrate this experimentally via simulations, collaboratively at that. In that regard, Dr. Conyers (2012) has given us some ideas for how to collaborate using AGP, summarized in Table 9.3. One submits we should heed this very helpful research framework for it has been offered by one of the more tireless practitioners of open-source mutual collaboration.

It will also be necessary continually to reflect back on many of the basic, major AGP questions that were posed above, and to develop many new ones. For example, how many different examples of nonanthropogenic subsurface material patterning can AGP techniques locate? Such materials can serve as considerable sources of confusion and can lead to expensive in-field mistakes if we do not recognize or learn to identify their nonanthropogenic nature. To answer the question "how many different examples of nonanthropogenic subsurface material patterning can AGP techniques locate?" we must turn to the literature as well as consult our own data, develop our own experience with interpreting AGP image models, and construct linking arguments, "arguments of relevance" (Fritz 1972:140), or "bridging arguments" (Smith 1977:611) to connect aspects of the archaeological record of interest to us to past causal phenomena (Wylie 2002). Further relevant questions to consider would be: How many legitimate

Table 9.3. Dr. Larry Conyers's Checklist for Successful Collaboration in AGP

1)	Communicate before, during, and after active AGP collaboration in the field.
2)	Get as much information as possible about physical conditions at study sites.
3)	Determine surficial vegetation conditions at study sites.
4)	Agree on survey area parameters among all collaborators.
5)	Bring a computer to the field and be willing to adjust survey area and data-collection parameters.
6)	Avoid overcollection of data at new sites. Spend some time calibrating equipment and people to how things "look" at sites.
7)	Tie AGP surveys to GPS or other known geographic and spatial data so surveys are not floating in space.
8)	Be willing to employ novel or ad hoc data-collection techniques should they prove necessary.
9)	Ensure that all collaborators have commensurate data, maps, and reports before leaving the field.
10)	Request digital copies of data and images and use extensive data backup procedures.
11)	Don't overinterpret limited data and be conservative in all interpretations.
12)	Distribute results widely and memorialize via the WWW, etc. (author's note: we need collaboratively to establish better open-source data-sharing protocols.*)

*All from Conyers 2012:204–207 except present author's note

examples of anthropogenic subsurface material can AGP "miss" and under which conditions? What human factors affect such results?

We should also ask ourselves what it is that we are trying to do. We should, again, consider AGP images of the subsurface as models of actual data, not photographs of them. To some extent, therefore, we are obligated to assess the plausibility of those image models of the subsurface through various means of testing, via additional AGP techniques or direct sampling. To link AGP data to causal phenomena, we can excavate sites, shovel- or augur-test them, core them, or we can establish as comprehensive as possible a comparative collection of AGP images of simulated features modeled on actual archaeological materials, using as wide as possible a variety of integrated techniques and sample spacings, or arrays. How useful would such resources by in assisting non-geophysical scholars to interpret geophysical results? A suitably comprehensive comparative collection, available in open-source format, would be an extremely powerful tool to assist in interpreting AGP data. Such a dedicated experimental program would also provide professionals and students at all levels of sophistication many opportunities to gain experience. Instead of involving a small community of geophysical purists, such an endeavor would represent a much broader cohort of collaborative parties.

What sorts of experiments could be done? That is rather a more difficult question to answer, but one can provide some basic ideas. For example, Leckebusch and Peikert (2001) used a brick in a sandbox to test ground-penetrating radar's three-dimensional capabilities. Enterprising graduate students could easily and rather cheaply construct similar sandboxes (perhaps 4 m^2 would be an ideal size) and test other simple materials with GPR, such as plastic milk jugs differentially filled with water, in full, half-full, quarter-full, and totally empty states. Similar experiments could be conducted using footballs, basketballs, and soccer balls, the idea being to test how things with variably-sized and shaped voids in them appear in AGP images. How do such materials appear to other AGP techniques? Have entire ranges of basketballs, milk jugs, cinder blocks, or anything been imaged geophysically in sand, in gravel, in silt/clay loam? Has anyone performed an analysis of how AGP techniques perform or don't in an artificial matrix of sidewalk or water-softener salt or salt mixed with soils or sand? Would archaeologists who work in salinated soils perhaps find it useful? What do ceramic roof tiles and field drainage tiles look like in AGP images? Would comprehensive experiments on imaging roof tiles and learning how (if possible) to distinguish roof tiles and drainage tiles from other buried materials perhaps be useful for classical archaeologists? How do cement or ceramic sculptures appear in AGP images? How does image resolution of all of the above vary as a function of different sediment matrices or matrix moisture content? The question to ask therefore really ought not to be what sorts of experiments could be done, but rather what experiments *cannot* be undertaken. Imagination is perhaps the only serious limit, provided one has access to AGP equipment.

One cannot say how receptive the "more traditional" community of AGP scholars and reviewers will be to the publication of such experimental research in, say, *Archaeological Prospection* or other professional journals. One suspects they might balk at publishing much of it, since it wouldn't have the advantage of portraying graphic images of actual archaeological materials. But those publication dynamics ought not to affect such basic and, one daresay necessary, research, and master's theses and doctoral dissertations, as well as conference posters and presentation papers would also be excellent places to engage in it. For we need to consider that while some experimental AGP research has been and is being conducted, very little of it has been theoretically *systematic* or geared toward actual anthropology, nor has it systematically explored wide ranges of variation in how different simulated materials and things appear differentially. Wide access to a systematic open-source database of such experimental and simulated research would be invaluable to all archaeologists seeking to use AGP in the field. For that matter, how difficult would it be to set up personal web pages, or social networking pages featuring such content? With enough interested parties making demands, we could also probably organize an actual online journal devoted purely to the publication of this and other crucial AGP research. Such actions might also stimulate the more mainstream or prestige journals to open their pages to publication of more pure experimental AGP research on simulations.

Conyers (2012) has indicated that archaeological deposits at Paleolithic and other ephemeral occupations can elicit images of many highly variable material configurations. Precisely because of that fact, experimentation on a wide range of simulated small-scale material configurations is absolutely crucial for providing us with good comparative data for recognizing real archaeological deposits at unexcavated sites or at sites with known unexcavated components. This volume should also therefore be seen as a call for broader implementation of AGP surveys at many pre-Neolithic open-air ephemeral sites. In that way, AGP data collected from real sites could be integrated with data from simulated experiments. We need to have both kinds of data available in order to have a truly representative and informative comparative collection at our disposal. Since it is unlikely that all sites will ever be found, and that fewer still will be excavated to any degree, we have few "real" indices of the real range of potential material patterning or how it will "look" in AGP data beyond experimental simulation and subsequent documentation.

A call for more AGP applications could also meet with some resistance from what Conyers (2012) terms "traditional archaeologists." A peculiar theme that one has heard expressed multiple times is that AGP will "put real diggers out of business if it gets good enough." While it is difficult to quantify "good enough," the sentiment does explain what one perceives as a real fear among some archaeologists in general, and CRM professionals in particular. To address such sentiments, one would just make several observations. First, regardless of the specific budgetary dynamics of archaeological excavations that vary on a site-by-site basis, in both academic and CRM venues, excavations are simply labor- and capital-intensive. Knowing where not to excavate through AGP survey is one way of avoiding wastes of labor and money. Now, there will be some professionals who are perfectly comfortable with not knowing what areas of a site to avoid expensive excavating for obvious reasons. This is not, however, a legitimate argument *against* widespread implementation of AGP surveys. Any AGP survey is likely to be less expensive than misdirecting what are today shrinking labor and capital investments in unproductive excavations. Yet, we should note that even the most exemplary AGP surveys do not reveal everything about a site. AGP will probably never develop to such an extent that the resolution of subsurface images will provide accurate models of everything buried at a site. In fact, until imaged material is excavated it really only "exists" in a state similar to Schrödinger's cat: it's nothing in particular until we reveal what it actually *is* through excavation and recovery. This means excavation is not really an endangered enterprise, nor are "traditional archaeologists" seriously imperiled by AGP, since excavation will still be necessary in many cases even after AGP survey.

The second point one would make would be that there are probably some sites that for various reasons are simply not suitable for AGP survey. In such cases, excavation will still remain the only viable means of actually testing their material content. Consider, for example, open-air ephemeral sites that are buried beneath thick overburdens. Those sorts of sites are quite common in many regions, such as the Magdalenian site of Pincevent in the Paris Basin of France, or multiple Paleoindian sites in the western United States. At sites like that there are probably no AGP techniques that are adequate to resolve small features of any configuration, which would obviate any AGP surveying. One suspects that pop culture might be to blame for this misimpression. There exists a very famous American motion picture about dinosaurs that have been cloned into existence from DNA drawn from amber-entrapped mosquitoes. One vignette from this movie presents a scenario in which paleontologists use seismic surveying of a dinosaur site to produce an extremely detailed video graphic of a dinosaur, including fine detail of a lunate wrist bone. Seismic surveying would never reveal such fine detail and neither would any currently available AGP technique. This just isn't possible yet, and probably will never be.

While it would certainly be wonderful if we could resolve AGP images of individual wrist bones or Clovis points, we certainly should not expect such abilities. Moreover, such abilities are not really the point of AGP techniques used in archaeology. Since we don't conduct excavations to resolve particular wrist bones or Clovis points we cannot make such a demand of AGP. Our AGP techniques, theory, and methodology will probably never be premised upon such minute specifics. Because of this consideration, excavations and archaeologists have nothing to fear and much to gain from employing them. For example, if an AGP-surveyed site features such a low density of archaeological deposits that it does not register in AGP data we will have learned something prior to excavations. We will have learned that the sites, or surveyed portions thereof, are not going to produce high densities of materials that show in AGP data. We will also probably have learned that the site will not produce much in the way of *patterned* artifactual density at all, or if it does it will be at very low densities and probably dispersed through various post-depositional processes. In other words, if the AGP equipment or the people using it didn't malfunction, the chances are very good that it works, and if nothing is "seen" then it follows that not much is there and traditional random subsurface survey techniques are appropriate. The results from the VBC 2009 excavations of the VBC 2006 radar survey certainly suggest this: at appropriate sites (those with homogeneous matrices that do not thwart the techniques), AGP can image artifacts if they are not ridiculously dispersed and spatially compromised. In sum, there are very few reasons *not* to apply AGP surveys at those sites where such techniques can be expected to function. Only in those instances where because of physical impediments they would be likely negated should AGP surveys be summarily dismissed.

In conclusion, however, very little of the hopeful collaboration and experimentation this volume advocates will come to fruition if the human agents able to accomplish them don't play their own parts. One fully realizes that they are requesting much of busy, accomplished professionals and students, all of whom have their own various research foci and entire hosts of individual constraints (not to mention their own lives). And yet, this is too big an undertaking for anyone to attempt it solo. The program described above *requires* collaboration, from the widest possible array of collaborators, at the widest possible variety of sites and site types, from the widest possible diversity of time periods. In fact, AGP presently suffers a bit from what could be termed a disciplinary *founder effect*, in which in which a minority of bottlenecked practitioners have passed a disproportionate amount of methodology and epistemology on to their descendents, such that the weight of tradition impinges somewhat upon further development. Anthropological archaeology and geophysical archaeology both must ask themselves, as disciplines and as individual archaeologists, how and what they intend AGP to be, within anthropology as a whole but also within the smaller confines of the field. If we don't ask more of AGP then what should we expect it to deliver? We should expect more of the same, which in publication is "we tried it over here; it worked/didn't work and here are or are not here the images we got from it." How many of us would be satisfied with that?

Bibliography

Abbas, A. M., T. F. Abdallatif, et al.

2005 Archaeological investigation of the eastern extensions of the Karnak Temple using ground-penetrating radar and magnetic tools. *Geoarchaeology* 20(5): 537–554.

Albert, Arthur

1963 A Mathematical Theory of Pattern Recognition. *The Annals of Mathematical Statistics.*

Aykroyd, R. et al.

2001 Bayesian Methods Applied to Survey Data from Archeological Magnetometry. *Journal of the American Statistical Association* 96(453):64–76.

Artz, Joseph.

2014 The Quaternary Geology of 13DB674, on the South Fork of Catfish Creek, Dubuque County, Iowa. EarthView Environmental Report on file with Tallgrass Historians, L.C.

Aspinall, A. et al.

2008 Editorial: *Archaeological Prospection*—The First Fifteen Years. *Archaeological Prospection* 15:241–245.

Audouze, F.

2010 Domesticity and Spatial Organization at Verberie. In *The Magdalenian Household: Unraveling Domesticity.* E. Zubrow, F. Audouze, J. Enloe (eds.) SUNY Press. pp. 145–175.

1987 The Paris Basin in Magdalenian Times. In, Olga Soffer (ed.), *The Pleistocene Old World: Regional Perspectives.* New York: Plenum Press. pp. 183–200.

Audouze, F., and J. G. Enloe

1997 High Resolution Archaeology at Verberie: Limits and Interpretations. *World Archaeology* 29(2):195–207.

Autsi, E.

2004 Ground-penetrating Radar Time-slices from North Ballachulish Moss. *Archaeological Prospection* 11:65–75.

Bevan, B.
2006 Geophysical Exploration for Buried Buildings. *Historical Archaeology* 40(4):27–50.

Bevan, B. W., and A. C. Roosevelt
2003 Geophysical exploration of Guajará, a prehistoric earth mound in Brazil. *Geoarchaeology* 18(3): 287–331.

Binford, L.
2001 Where Do Research Problems Come From? *American Antiquity* 66(4): 669–678.
1984 Butchering, Sharing, and the Archaeological Record. *Journal of Anthropological Archaeology* 3:235–257.
1983 *In Pursuit of the Past: Decoding the Archaeological Record*: with a New Afterword. University of California Press.
1980 Willow Smoke and Dogs' Tails: Hunter-Gatherer Settlement Systems and Archaeological Site Formation. *American Antiquity* 45(1):4–20.
1981 *Bones: Ancient Men and Modern Myths.* Academic Press.
1977 Introduction. In *For Theory Building in Archaeology: Essays on Faunal Remains, Aquaic resources, Spatial Analysis, and Systemic Modeling.* L. Binford (ed.) pp. 1–10. New York: Academic Press.
1968 Post-Pleistocene Adaptations. In *New Perspectives in Archaeology,* S. R. Binford and L. R. Binford (eds.), 313–341. Chicago: Aldine.
1967 Smudge Pits and Hide Smoking: The Use of Analogy in Archaeological Reasoning. *American Antiquity* 32(1):1–12.
1962 Archaeology as Anthropology. *American Antiquity* 28(2):217–225.

Binford, L., and J. Bertram
1977 Bone Frequencies and Attritional Processes. In: Binford (ed.) *For theory Building in Archaeology: Essays of Faunal Remains, Aquatic Resources, Spatial Analyses, and Systemic Modeling,* pp. 77–153. Academic Press.

Blau, S., et al.
2000 Seeing Through the Dunes: Geophysical Investigations at Muweilah, an Iron Age Site in the United Arab Emirates. *Journal of Field Archaeology* 27(2):117–129.

Bodu, P.
2010 Espaces et Habitats au Tardiglaciaire dans le Bassin Parisien: Une Illustration avec les Gisements Magdalenian de Pincevent et Azilien du Closeau. In *The Magdalenian Household: Unraveling Domesticity.* E. Zubrow, F. Audouze, J. Enloe (eds.) SUNY Press. pp. 176–197.

Bogucki, P.
1985 Theoretical directions in European archaeology. *American Antiquity* 50(4):780–788.

Boniger, U., and J. Tronicke

2010 Improving the Interpretability of 3D GPR Data Using Target-Specific Attributes: Applications to Tomb Detection. *Journal of Archaeological Science* 37(2):360–367.

Booth, A., N. Linford, R. Clark, and T. Murray

2008 Three-Dimensional, Multi-Offset Ground-penetrating Radar Imaging of Archaeological Targets. *Archaeological Prospection* 15:93–112.

Bourquin, S., J. Vairon, P. Le Strat

1997 Three-Dimensional Evolution of the Paris Basin Based on Detailed Isopach Maps of the Stratigraphic Cycles: Tectonic Influences. *Geology Rundsch* 86:670–685.

Carrozzo, M., G. Leucci, S. Negri, and L. Nuzzo

2003 GPR Survey to Understand the Stratigraphy of the Roman Ships Archaeological Site (Pisa, Italy). *Archaeological Prospection* 10:57–72.

Chadwick, W. J., and J. A. Madsen

2000 The application of ground-penetrating radar to a coastal prehistoric archaeological site, Cape Henlopen, Delaware, USA. *Geoarchaeology* 15(8): 765–781.

Chapman, H., J. Adcock, and J. Gater

2009 An approach to mapping buried prehistoric palaeosols of the Atlantic seaboard in Northwest Europe using GPR, geoarchaeology and GIS and the implications for heritage management. *Journal of Archaeological Science* 36:2308–2313.

Chazan, M.

2011 *World Prehistory and Archaeology: Pathways through time.* Pearson Press. Chianese, D., M. D'Emilio, S. Di Salvia, V. Lapenna, M. Ragosta and E. Rizzo

2004 Magnetic mapping, ground penetrating radar surveys and magnetic susceptibility measurements for the study of the archaeological site of Serra di Vaglio (southern Italy). *Journal of Archaeological Science* 31 (5):633–643.

Collins, J., and B. Molyneaux

2003 *Archaeological Survey.* Rowman Altamira Press.

Conyers, L.

2013 *Ground-penetrating radar for Archaeology Third Edition.* AltaMira Press.

2012 *Interpreting Ground-penetrating Radar for Archaeology.* Left Coast Press.

2010 Ground-penetrating radar for anthropological research. *Antiquity* 84(323):175–184.

2006a Ground-Penetrating Radar Techniques to Discover and Map Historic Graves. *Historical Archaeology*, 40(3):64–73.

2006b Innovative Ground-Penetrating Radar Methods for Archaeological Mapping. *Archaeological Prospection* 13:139–141.

2004 *Ground-penetrating radar for Archaeology.* AltaMira Press.
1995 The use of ground-penetrating radar to map the buried structures and landscape of the Ceren site, El Salvador. *Geoarchaeology* 10(4): 275–299.

Conyers, L., and C. Cameron
1998 Ground-Penetrating Radar Techniques and Three-Dimensional Computer Mapping in the American Southwest. *Journal of Field Archaeology* 25(4):417–430.

Conyers, L., J. Michael Daniels, Jonathan A. Haws and Michael M. Benedetti.
2013 An Upper Palaeolithic Landscape Analysis of Coastal Portugal Using Ground-penetrating Radar. *Archaeological Prospection* 20:45–51.

Conyers, L., E. Ernenwein, M. Grealy, and K. Lowe
2008 Electromagnetic Conductivity Mapping for Site Prediction in Meandering River Floodplains. *Archaeological Prospection* 15:81–91.

Conyers, L. B., and C. Gaffney
2000 Some Reflections on the Society for American Archaeology Annual Meeting. *Archaeological Prospection* 7:145–146.

Conyers, L., and D. Goodman
1997 *Ground-penetrating Radar: An Introduction for Archaeologists.* Altamira Press: London.

Conyers, L., and J. Leckebusch
2010 Geophysical archaeology research agendas for the future: Some ground-penetrating radar examples. *Archaeological Prospection* 17:117–123.

Cordier, S., D. Harmand, M. Frechen, and M. Beiner
2005 Fluvial system response to Middle and Upper Pleistocene climate change in the Meurthe and Moselle valleys (Eastern Paris Basin and Rhenish Massif). *Quaternary Science Reviews* 25:1460–1474.

Crapo, R. H.
2013 *Cultural Anthropology*. Bridgepoint Education.

Diamond, J.
1999 *Guns, Germs, and Steel: the Fates of Human Societies.* New York: Norton Publishers.

Dionne, C., D. Wardlaw, and J. Schultz
2010 Delineation and Resolution of Cemetery Graves Using a Conductivity Meter and Ground-Penetrating Radar. *Technical Briefs in Historic Archaeology* 5:20–30.

Dogan, M., and S. Papamarinopoulos
2006 Exploration of the Hellenistic Fortification Complex at Asea Using a Multigeophysical Prospection Approach. *Archaeological Prospection* 13, 1–9.

Due Trier, O., S. Larsen, and R. Solberg
2009 Automatic Detection of Circular Structures in High-resolution Satellite Images of Agricultural Land. *Archaeological Prospection* 16:1–15.

Dunnell, R.
1982 The Harvey Lecture Series. Science, Social Science, and Common Sense: The Agonizing Dilemma of Modern Archaeology. *Journal of Archaeological Science* 38(1):1–25.

Enloe, James
2010 Technology and Demographics: An Introduction. In *The Magdalenian Household: Unraveling Domesticity.* E. Zubrow, F. Audouze, J. Enloe (eds.) SUNY Press. pp. 11–14.
2006 Geological Processes and Site Structure: Assessing Integrity at a Late Paleolithic Open-Air Site in Northern France. *Geoarchaeology* 21(6):523–540.
2004 Equifinality, Assemblage Integrity and Behavioral Inferences at Verberie. *Journal of Taphonomy* 2(3):147–165.
1993 Subsistence Organization in the Early Upper Paleolithic: Reindeer Hunters of the Abri du Flageolet, Couche V. In *Before Lascaux: the Complex Record of the Early Upper Paleolithic,* H Knecht, A. Pike-Tay, R. White (eds.), pp. 101–115. CRC Press.

Enloe, J., and F. Audouze
2010 The Magdalenian site of Verberie (Le Buisson Campin): An Overview. In *The Magdalenian Household: Unraveling Domesticity.* E. Zubrow, F. Audouze, J. Enloe (eds.) SUNY Press. pp. 15–21.

Enloe, J., Storey, G., Thompson, J., and Audouze F.
2007 Ground Penetrating Radar in the Paleolithic: A Survey of an Open air site at Verberie, France. Paper presented in session "Remote Sensing and Geophysical Techniques," at the 72nd annual meeting of the Society for American Archaeology, Austin, TX, April 25–29, 2007.

Fagan, Brian
2005 *Ancient North America: the Archaeology of a Continent.* Thames & Hudson.

Flannery, K.
1982 The Golden Marshalltown: A Parable for the Archeology of the 1980s. *American Anthropologist, New Series* 84(2):265–278.

Fontijn, D.
2007 The Significance of 'Invisible' Places. *World Archaeology* 39(1):70–83.

Forte, E., and M. Pipan
2008 Integrated Seismic Tomography and GPR for the High-Resolution Study of Burial Mounds (tumuli). *Journal of Archaeological Science* 35(9): 2614–2623.

Foxhall, L.
2000 The Running Sands of Time: Archaeology and the Short-Term. *World Archaeology* 31(3):484–498.

Fritz, J. M.
1972 Archaeological Systems for Indirect Observation of the Past. In M. P. Leone (ed.) *Contemporary Archaeology*. pp. 135–157. Carbondale: Southern Illinois University Press.

Gaffney, C., and J. Gater
2003 *Revealing the buried past: Geophysics for Archaeologists*. Tempus Publishers.
1999 Popularizing Archaeological Geophysics: The "Time Team" Experience on British Television. In *Archaeological Prospection: Third International Conference on Archaeological Prospection*, edited by J. W. E. Fassbinder and W. E. Irlinger, p. 35. Bayerisches Landesamt fir Denkmalpflege, Munich.

Gaffney, C., J. Gater, T. Saunders, J. Adcock
2004 D-Day: Geophysical Investigation of a World War II German Site in Normandy, France. *Archaeological Prospection* 11:121–128.

Gamble, Clive
2001 *Archaeology: the Basics*. Taylor and Francis.

Garrison, E.
1996 Archaeogeophysical and Geochemical Studies at George Washington Carver National Monument, Diamond, Missouri. *Historical Archaeology* 30(2):22–40.

Garrison, et al.
1985 Magnetic Prospection and the Discovery of Mission Santa Catalina de Guale, Georgia. *Journal of Field Archaeology* 12(3):299–313.

Gibbon, Guy
1984 *Anthropological Archaeology*. Columbia University Press.

Goodman, D., Y. Nishimura, H. Hongo, and N. Higashi
2006 Correcting for Topography and the Tilt of GPR Antennae. *Archaeological Prospection* 13:157–161.

Goodyear, A., L. M. Raab, T. Klinger
1978 The Status of Archaeological Research Design in Cultural Resource Management. *American Antiquity* 43(2):159–173.

Grasmueck, M., R. Weger, and L. Hortsmeyer
2004 Three-Dimensional Ground-Penetrating Radar Imaging of Sedimentary Structures, Fractures, and Archaeological Features at Submeter Resolution. *Geology* 11/04 21(11):933–936.

Grealy, M.
2006 Resolution of GPR Reflections at Differing Frequencies. *Archaeological Prospection* 13:142–146.

Hardy-Smith, T., and P. Edwards
2004 The Garbage Crisis in prehistory: artefact discard patterns at the Early Natufian site of Wadi Hammeh 27 and the origins of household refuse disposal strategies. *Journal of Anthropological Archaeology* 23:253–289.

Hargrave, M. et al.
2007 Magnetic Evidence of Ridge Construction and Use at Poverty Point. *American Antiquity* 72(4):757–769.

Hayden, B.
1993 *Archaeology: the Science of Once and Future Things.* W.H. Freeman and Company.

Hietala, H., and D. Stevens
1977 Spatial Analysis: Multiple Procedures in Pattern Recognition Studies. *American Antiquity* 42(4):539–559.

Hildebrand, J., S. Wiggins, P. Henkart, and L. Conyers
2002 Comparison of Seismic Reflection and GPR Imaging at the Controlled Archaeological Test Site, Champaign, Illinois. *Archaeological Prospection* 9:9–21.

Hill, C.
2001 Geologic contexts of the Acheulean (Middle Pleistocene) in Eastern Sahara. *Geoarchaeology* 16(1):65–94.

Hodgetts, L. et al.
2011 Archaeological magnetometry in an Arctic setting: a case study from Maguse Lake, Nunavut. *Journal of Archaeological Science* 38 (2011) 1754–1762.

Hurst Thomas, D. et al.
2011 Life among the tides: recent archaeology on the Georgia Bight: proceedings of the sixth Caldwell Conference, St. Catherine's Island, Georgia, May 20–22, 2011. Anthropological papers of the American Museum of Natural History, No. 98. URL: http://digitallibrary.amnh.org/dspace/handle/2246/6435.

Isayev, E.
2006 Archaeology ≠ Object as History ≠ Text: nudging the special relationship into the post-ironic. *World Archaeology* 38(4):599–610.

Johnson, M.
2006 On the nature of theoretical archaeology and archaeological theory. *Archaeological Dialogues* 13(2):115–116.

Jones, G., and G. Munson

2005 Geophysical Survey as an Approach to the Ephemeral Campsite Problem: Case Studies from the Northern Plains. *Plains Anthropologist* 50(193):31–43.

Kenyon, J.

1977 Ground-Penetrating Radar and Its Application to a Historical Archaeological Site. *Historical Archaeology* 11:48–55

Klasner, J., and P. Calengas

1981 Electrical Resistivity and Soil Studies at Orendorf Archaeological Site, Illinois: A Case Study. *Journal of Field Archaeology* 8(2):167–174.

Kosso, P.

1991 Empirical epistemology and philosophy of science. *Metaphilosophy* 22(4):349–363.

Kvamme, K.

2008 "Remote Sensing Approaches to Archaeological Reasoning: Pattern Recognition and Physical Principles." *Archaeological Concepts for the Study of the Cultural Past* (2008): 65.

2003 Geophysical Surveys as Landscape Archaeology. *American Antiquity* 68(3):435–457.

Kvamme, K., and S. Ahler

2007 Integrated Remote Sensing and Excavation at Double Ditch State Historic Site, North Dakota. *American Antiquity* 72(3):539–561.

Leckebusch, J.

2005 Precision Real-time Positioning for Fast Geophysical Prospection. *Archaeological Prospection* 12:199–202.

2003 3D GPR: A Modern Three-Dimensional Prospection Method. *Archaeological Prospection* 10:213–240.

Leckebusch, J., and R. Peikert

2001 Investigating the True Resolution and Capabilities of Ground-Penetrating Radar in Archaeological Surveys: Measurements in a Sand Box. *Archaeological Prospection* 8:29–40.

Leucci, Giovanni

2006 Contribution of ground-penetrating radar and electrical resistivity tomography to identify the cavity and fractures under the main Church in Botrugno (Lecce, Italy). *Journal of Archaeological Science* 33(9):1194–1204.

Linington, R.

1963 The Application of Geophysics to Archaeology. *American Scientist* 51(1):48–70.

Lorenzo, H., and P. Arias

2005 A Methodology for Rapid Archaeological Site Documentation Using Ground-Penetrating Radar and Terrestrial Photogrammetry. In *Geoarchaeology: An International Journal* 20 (5):521–535.

Mahar, G.
2011 Archaeological Geophysics as Primary Data. Paper presented at the 76th Annual Meeting of the Society for American Archaeology, Sacramento, California.

Martinez, A., and O. Hamsici
2008 Who is LB1? Discriminant analysis for the classification of specimens. *Pattern Recognition* 41:3436–3441

McGuire, R.
2008 *Archaeology as Political Action.* California Series in Public Anthropology.

McNutt, C., J. Franklin, and E. Henry
2012 New Perspectives on Mississippian Occupations in Western Tennessee and Northwestern Mississippi: Recent Chronological and Geophysical Investigations. *Southeastern Archaeology* 31:231–250.

Moorman, B., S. Robinson, and M. Burgess
2003 Imaging Periglacial Conditions with Ground-penetrating Radar. In *Permafrost and Periglacial Processes* 14:319–329.

Mussett, A., and M. Khan
2000 *Looking into the Earth: An Introduction to Geological Geophysics.* Cambridge University Press.

Nanda, S., and R. L. Warms
2007 *Cultural Anthropology.* Cengage Learning.

Negri, S., and Leucci, G.
2006 Geophysical investigation of the Temple of Apollo (Hierapolis, Turkey): *Journal of Archaeological Science* 33:1505–1513.

Neubauer et al.
2002 First deep seismic reflection images of the Eastern Alps reveal giant crustal wedges and transcrustal ramps *Geophysical Research Letters* 29(10).

Odell, G.
1996 Micro-wear in Perspective: A Sympathetic Response to Lawrence H. Keeley. *World Archaeology* 7:226–240.

Oms, et al.
2000 Early human occupation of western Europe: Paleomagnetic dates for two paleolithic sites in Spain. *PNAS,* 97(19):10666–10670.

Ozawa, K.
1990 Simulation studies of optical illusions based on a position dependent point-spread function. *Pattern Recognition* 23(12):1361–1306.

Park, Michael
2008 *Introducing Anthropology—An Integrated Approach.* McGraw-Hill Humanities/Social Sciences/Languages.

Pastre, J. F., N. Limondin-Lozouet, C., Leroyer, P. Ponel, and M. Fontugne
2002 River system evolution and environmental changes during the Late glacial in the Paris Basin (France). *Quaternary Science Reviews* 22:2177–2188.

Patch, S., and D. Gregory
2010 Refining Middle Woodland Site Structure: Ground Penetrating Radar Investigations at Site 40Mi70, Marion County, Tennessee. Paper presented in session "Anthropological Geophysics: Scale and configuration in the archaeological record" at 74th Annual Society for American Archaeology Meeting. St. Louis. 2010.

Piro, S., P. Mauriello, F. Cammarano
2000 Quantitative Integration of Geophysical Methods for Archaeological Prospection. *Archaeological Prospection* 7:203–213.

Pomfret, J.
2006 Ground-penetrating Radar Profile Spacing and Orientation for Subsurface Resolution of Linear Features. *Archaeological Prospection* 13:151–153.

Quarto, R. et al.
2007 Ground penetrating radar survey of a prehistoric site in Southern Italy. *Journal of Archaeological Science* 34:2071–2080.

Raab, L., and A. Goodyear
1984 Middle-range theory in archaeology: a critical review of origins and applications. *American Antiquity* 49(2):255–268.

Renfrew, C. and P. Bahn
1991 *Archaeology: Theories, Methods and Practice.* Thames and Hudson. Rodrigues, S., J. Porsani, V. Santos, P. DeBlasis, P. Giannini
2009 GPR and inductive electromagnetic surveys applied in three coastal sambaqui (shell mounds) archaeological sites in Santa Catarina state, South Brazil. *Journal of Archaeological Science* 36:2081–2088.

Rogers, L.
2008 Dubuque Southwest Arterial Highway Project Alternative E—Design Variation 1: Archaeological Data Recovery Plan for Sites in APE. CRM Project proposal available online at: http://ia-dubuque.civicplus.com/DocumentCenter/View/15365

Rosenthal, G.
2007 Spatiotemporal Dimensions of Visual Signals in Animal Communication. *Annual Review of Ecology, Evolution, and Systematics* 38:155–178.

Rowlands, A., and A. Sarris
2007 Detection of Exposed and Subsurface Archaeological Remains using Multi-Sensor Remote Sensing. *Journal of Archaeological Science* 34(5):795–803.

Ruffell, A. et al.

2004 Ground-penetrating Radar Facies as an aid to Sequence Stratigraphic Analysis: Application to the Archaeology of Clonmacnoise Castle, Ireland. *Archaeological Prospection* 11:247–262.

Salmon, M.

1975 Confirmation and explanation in archaeology. *American Antiquity* 40(4):459–464.

Schiffer, M.

1987 *Formation Processes of the Archaeological Record.* Albuquerque: UNM Press. *Behavioral Archaeology*. Academic Press.

1972 Archaeological context and systemic context. *American Antiquity* 37(2):156–165.

Scultz, J. et al.

2006 Sequential Monitoring of Burials Containing Large Pig Cadavers Using Ground-Penetrating Radar. *Journal of Forensic Science*, 51 (3).

Smith, B. D.

1977 Archaeological Inference and Inductive Confirmation. *American Anthropologist* 79:598–617.

Solla, M, B. Riveiro, M. Alvarez, and P. Arias

2012 Experimental forensic scenes for the characterization of ground-penetrating radar wave response. *Forensic Science International* 220(1–13):1–9.

South, Stanley

1978 Pattern Recognition in Historical Archaeology. *American Antiquity* 3(2):223 230.

Speth, J.

2004 News flash: negative evidence convicts Neanderthals of gross mental incompetence. *World Archaeology* 36(4):519–526

Speth, J., and K. Spielman

1983 Energy Source, Protein Metabolism, and Hunter-Gatherer Subsistence Strategies. *Journal of Anthropological Archaeology* 2(1):1–31.

Storey, G. R., and W. E. Whittaker

2005a Ground-Penetrating Radar Survey of the Effigy Mounds National Monument Sny Magill Mound Group (13CT18), Clayton County. Contract Completion Report 1233. Office of the State Archaeologist, Iowa City. 2005.

2005b Ground-Penetrating Radar Survey of the Possible 13AM446, Effigy Mounds National Monument, Allamakee County, Iowa. Completion Report 1234. Office of the State Archaeologist, Iowa City. 2005.

2005c Ground Penetrating Radar Survey of the Sny Magill Mound Group, Effigy Mounds National Monument, Iowa. Under review for *American Antiquity*.

2004a Ground-Penetrating Radar Survey for Primary Roads Project NH5-151-4(84)—3H-53 of Portions of the Bowen's Prairie Cemetery (13JN152), Section 6, T86N-R2W, Jones County, Iowa. Contract Completion Report 1110. Office of the State Archaeologist, University of Iowa, Iowa City.

2004b Ground-Penetrating Radar Survey of the Herbert Hoover Boyhood Home (13CD134), West Branch, Cedar County, Iowa (with William E. Whittaker). Contract Completion Report 1123. Office of the State Archaeologist, Iowa City.

Tang, G., and T. Huang

1980 Using the Creation Machine to Locate airplanes on aerial photos. *Pattern Recognition* 12:431–442.

Thacker, P., and B. Elwood

2002 Detecting Palaeolithic Activity Areas Through Electrical Resistivity Survey: An Assessment from Vale de Obidos, Portugal. *Journal of Archaeological Science* 29, 563–570.

Thomas, H.

1987 Seismic hazards on the Cascadia subduction zone, *Science* 236:162–168.

Thompson, Jason

2014a *A Composite View to the Past: A Methodological Integration of Zooarchaeology and Geophysics at the Magdalenian Site of Verberie le Buisson-Campin.* British Archaeological Reviews/Archaeopress.

2014b Ground-penetrating radar Failure at 13DB674 in Iowa or redefinition of success? Personal report on file with Tallgrass Historians, L.C.

2012 The allegory of the hammer and the nail gun and other unstable orthodoxies of "modernity": possible pitfalls of "behavioural modernity." *Newsletter of the Australian Rock Art Research Association* 29(2):3–12.

2011 Views to the Past: Faunal and Geophysical Analysis of the Open-Air Upper Paleolithic Site of Verberie. Unpublished Doctoral Dissertation on file at The University of Iowa and on ProQuest and UMI.

Thompson, J., and G. Storey

n.d. Ground-Penetrating Radar and Imaging of Complex Subsurface Archaeological Materials: recreating the French Upper Paleolithic. In peer review at *Archaeological Prospection*

2010 Ground-Penetrating Radar and Imaging of Complex Subsurface Archaeological Materials. Paper presented in session "Anthropological Geophysics: Scale and configuration in the archaeological record" at 74th Annual Society for American Archaeology Meeting. St. Louis. 2010.

Thompson, Victor D., Phillip J. Arnold, III, Thomas, J. Pluckhahn, and Amber VanDerwarker
2011 Situating Remote Sensing in Anthropological Archaeology. *Archaeological Prospection* 18(3): 195–213.

Trigger, Bruce
1998 *Sociocultural Evolution: Calculation and Contingency.* Blackwell Press.
1989 *A History of Archaeological Thought.* Cambridge University Press.

Vafidis, A., N. Economou, Y. Ganiatsos, M. Manakou, G. Poulioudis, G. Sourlas, E. Vrontaki, A. Sarris, M. Guy, and Th. Kalpaxis
2005 Integrated geophysical studies at ancient Itanos (Greece). *Journal of Archaeological Science* 32(7):1023–1036.

Watson, Patty Jo
1991 A Parochial Primer: The New Dissonance as Seen from the Midcontinental USA. In *Processual and Postprocessual Archaeologies: Multiple Ways of Knowing the Past.* Preucel (ed.), pp. 265–274. Southern Illinois University Press, Carbondale.

Weaver, W.
2006 Ground-penetrating Radar Mapping in Clay: Success from South Carolina, USA. *Archaeoogical Prospection* 13:147–150.

Weinstein-Evron, M. et al.
2002 Geophysical investigations in the service of Mount Carmel (Israel) prehistoric research. *Journal of Archaeological Science* 30 (2003) 1331–1341.

Wheeler, J.
2007 Geophysics in the Garden: A Survey of the Gardens at Duncombe Park, North Yorkshire. *Garden History* 35(1):85–91.

Wheeler, J. A. and W. H. Zurek, eds.
1983 *Quantum Theory and Measurement.* Princeton University Press, New Jersey.

Willig, M. R., D. M. Kaufmann, and R. D. Stevens
2003 Latitudinal gradients of biodiversity: pattern, process, scale and synthesis. *Annu. Rev. Ecol. Syst.* 34:273–309.

Wilson, G.
2010 Community, Identity, and Social Memory at Moundville. *American Antiquity* 75(1):3–18

Witten, A.
2006 *Handbook of geophysics and archaeology.* Equinox Publishing.

Wylie, A.
2002 *Thinking from Things : Essays in the Philosophy of Archaeology.* Berkeley: University of California Press.
1985 Between philosophy and archaeology. *American Antiquity* 50(2):478–490.

Wurtele, Z.

1965 A Problem in Pattern Recognition. *Journal of the Society of Industrial and Applied Mathematics* 13(1).

Yalciner, C. et al.

2009 New Temple Discovery at the Archaeological Site of Nysa (Western Turkey) using GPR Method. *Journal of Archaeological Science* 36(8):1680–1689.

Zhou, H., and M. Sato

2001 Archaeological investigation in Sendai Castle using ground-penetrating Radar. *Archaeological Prospection* 8:1–11.

Index

About the Author

Jason Randall Thompson is an anthropological archaeologist who has worked in the Middle and Upper Paleolithic of France (Solutré and Verberie le Buisson-Campain), the Middle and Upper Paleolithic of Jordan (Ain al-Buhayra and Thalab al-Buhayra), and the Midwestern United States. Jason currently teaches North American prehistory/archaeology and Human Origins at University of Northern Iowa, Anthropology at Ashford University, and World Archaeology, Physical Anthropology, Forensic Anthropology, Cultural Anthropology, and Sociology at American Public University. Jason's research includes Paleolithic Archaeology of Eurasia, the Archaic period of North America, anthropological archaeology, geophysical archaeology, faunal analysis, lithic analysis and lithic replication studies, Human Origins, Archaic behavioral modernity, and experimental replication of perishable Paleolithic technologies.

www.ingramcontent.com/pod-product-compliance
Ingram Content Group UK Ltd.
Pitfield, Milton Keynes, MK11 3LW, UK
UKHW041122071026
15587UKWH00009B/95